Synthesizing Materials in Microgravity

Synthesizing Materials in Microgravity: Unlocking Novel Materials Beyond Earth's Limits reveals how space-based environments unlock breakthroughs in material design by eliminating Earth's gravitational constraints. On Earth, gravity induces convection, sedimentation, and pressure gradients that compromise material quality during crystallization, solidification, polymerization, and sintering. In microgravity, such as aboard the ISS, these forces are minimized, enabling precise control of atomic arrangements, phase transitions, and microstructure formation. The result: materials with superior structural, electrical, and thermal properties once thought unattainable.

This book presents cutting-edge experiments from space, bridging theoretical foundations with real-world case studies across alloys, ceramics, semiconductors, and polymers. With high-quality visuals, it demystifies complex phenomena and highlights future commercial and scientific frontiers. A vital resource for scientists, engineers, and innovators seeking new material horizons.

Off-world Science & Engineering Series

Michael D. Holloway

A 7-book series to explore the transformative role of microgravity across disciplines such as materials science, biology, engineering, cognition, and policy unveiling how humanity can transition from Earth-bound limitations to sustainable off-world civilizations. This series connects cutting-edge research with visionary applications, offering a roadmap for those engineering the future of life in space.

Effects of Microgravity on Terrestrial Born Biological Systems
Michael D. Holloway

Synthesizing Materials in Microgravity: Unlocking Novel Materials Beyond Earth's Limits
Michael D. Holloway

Synthesizing Materials in Microgravity

Unlocking Novel Materials Beyond Earth's Limits

Michael D. Holloway

CRC Press
Taylor & Francis Group
Boca Raton New York London

CRC Press is an imprint of the
Taylor & Francis Group, an **informa** business

First edition published 2026
by CRC Press
2385 NW Executive Center Drive, Suite 320, Boca Raton FL 33431

and by CRC Press
4 Park Square, Milton Park, Abingdon, Oxon, OX14 4RN

CRC Press is an imprint of Taylor & Francis Group, LLC

ISBN: 978-1-041-10636-4 (hbk)
ISBN: 978-1-041-10637-1 (pbk)
ISBN: 978-1-003-65608-1 (ebk)

DOI: 10.1201/9781003656081

Typeset in Minion
by SPi Technologies India Pvt Ltd (Straive)

This work is dedicated to Fran Christopher for keeping things grounded while looking beyond.

Contents

Preface

Maybe we will be able to synthesize self-replicating molecules capable of self-replication, catalysis, and evolution within an appropriate environment. We haven't been able to do so on Earth, perhaps due to the constraints of gravity?

Making amazing stuff in space is gravy, just have to take 'it' out...

MDH

About the Author

President of 5th Order Industry, which provides training, failure analysis, and designed experiments, with 40 years' experience in industry starting with research and product development for Olin Chemical, WR Grace, Rohm and Haas, and GE Plastics reliability engineering and analysis for NCH, ALS, and SGS, a subject matter expert in failure analysis, reliability engineering, and designed experiments for science and engineering, I hold 16 professional certifications, a patent, a MS Polymer Engineering, BS Chemistry, BA Philosophy, authored 12 books, contributed to several others, have been cited in over 1,000 manuscripts and several hundred master's theses and doctoral dissertations.

Acknowledgements

I'd like to thank Garth Battista, for his friendship and for the use of the image on the cover. If he hadn't kicked my prom date and me out of Harkness Memorial State Park, we never would have ended up at Pleasure Beach, where the photo was taken. Sometimes, impatient judgment leads to perfect outcomes.

To Jana Stoudemire, for humoring me years ago as I rambled on about my wild ideas - your patience helped launch this whole thing (pun very much intended).

To Scott Joseph Kelly, who probably doesn't remember me, but once asked why I never became a Mission Specialist:

"Sir, space is a great place to make stuff, but it's no place for a human to live!"

And finally, to Fran Christopher, for keeping me grounded on Earth while still helping me launch ideas into orbit.

Introduction

Material synthesis on Earth is inherently constrained by gravity, which exerts a profound influence on processes such as crystallization, alloy solidification, polymerization, and ceramic sintering. Gravitational forces drive buoyancy-induced convection, sedimentation, and hydrostatic pressure gradients, which often result in defects, non-uniform microstructures, and phase segregation in synthesized materials. These imperfections limit the performance and applicability of advanced materials in fields like aerospace engineering, electronics, and biomedical devices. Researchers have long sought to overcome these gravitational limitations through precise control over temperature gradients, chemical compositions, and process parameters. However, terrestrial experiments continue to face intrinsic physical barriers. The advent of space-based research platforms, such as the International Space Station (ISS), has introduced a transformative opportunity: material synthesis in microgravity. In this unique environment, gravitational forces are largely eliminated, allowing for unprecedented control over atomic arrangement, phase distribution, and crystallization dynamics, leading to materials with unattainable properties on Earth.

CHAPTER 1

The Role of Gravity in Material Synthesis

On Earth, gravitational forces impact material synthesis at both microscopic and macroscopic scales. In crystallization processes, heavier atoms or phases tend to settle while lighter ones rise, disrupting uniform crystal growth. Similarly, buoyancy-driven convection in molten metals and polymer solutions creates uneven thermal and chemical gradients, leading to irregular structures and defects. Hydrostatic pressure gradients further alter the equilibrium conditions during crystallization, polymerization, and sintering, preventing precise control over microstructure formation. While terrestrial experiments attempt to mitigate these effects using carefully designed thermal profiles, vibration isolation systems, and chemically balanced growth mediums, gravity remains an insurmountable constraint for certain high-precision applications.

MICROGRAVITY: A UNIQUE PLATFORM FOR MATERIAL INNOVATION

Microgravity environments fundamentally alter the dynamics of material synthesis by minimizing buoyancy, sedimentation, and convective currents. In microgravity, diffusion, the random movement of particles, is the dominant mode of mass transport, leading to significantly more uniform mixing and distribution of elements and phases at the molecular level. The absence of sedimentation allows suspended particles or molecules to remain evenly distributed throughout the synthesis process,

DOI: 10.1201/9781003656081-1

enabling materials to form without gravitation-induced imperfections. Furthermore, the reduction of hydrostatic pressure gradients eliminates stress differentials during cooling, crystallization, and solidification, allowing for highly ordered atomic arrangements and defect-free structures. These advantages have unlocked pathways for creating materials with unique properties, including enhanced strength, purity, optical clarity, and functional precision.

Unique Crystal Syntheses in Microgravity

One of the most exciting opportunities presented by microgravity research lies in unique crystal synthesis. On Earth, crystal growth is often marred by convective currents, uneven thermal gradients, and gravitational settling, resulting in inclusions, dislocations, and irregularities. In contrast, microgravity facilitates the growth of near-perfect single crystals, essential for applications in semiconductors, optics, and pharmaceuticals. For example, protein crystals synthesized in microgravity exhibit significantly higher levels of structural order and size compared to their terrestrial counterparts, improving their utility in drug design and medical treatments.

Another groundbreaking advancement is the synthesis of *quasicrystals*, materials with ordered yet non-repeating atomic patterns. These structures, first discovered on Earth, have shown improved purity and stability when grown in microgravity due to the absence of gravitational disruption during atomic arrangement. Similarly, perovskite crystals, essential for next-generation solar cells, have been synthesized in space with enhanced electronic properties and fewer structural defects, offering superior performance compared to Earth-grown samples.

Synthesis of Organic Materials in Microgravity

Organic materials, including proteins, pharmaceuticals, and organic semiconductors, have demonstrated exceptional improvements when synthesized in microgravity. Protein crystallization, in particular, benefits from microgravity's stable growth conditions, resulting in larger and more uniform crystals. This impacts drug design directly, as the structural resolution of protein crystals is critical for developing targeted therapies. Similarly, organic thin films and nanostructures, used in flexible electronics and organic light-emitting diodes (OLEDs), exhibit improved structural uniformity and defect-free interfaces when deposited in microgravity conditions.

Ceramic Materials: Precision Beyond Gravity

Ceramics are valued for their exceptional mechanical strength, thermal stability, and electrical insulation properties. However, synthesizing defect-free ceramics on Earth is hindered by gravitational forces, which cause sedimentation and phase separation during sintering and crystallization. In microgravity, ceramic particles remain uniformly dispersed in molten or semi-molten states, resulting in dense, homogeneous microstructures with minimal inclusions. For instance, transparent ceramic materials used in high-performance optical lenses and advanced photonics have been synthesized in microgravity with unparalleled clarity and defect-free surfaces. Additionally, microgravity allows for the development of functionally graded ceramics (FGCs), where material properties change seamlessly across different regions, a feat challenging to achieve under terrestrial gravity.

Metallic Alloys: Homogeneous and Defect-Free Structures

Metallic alloys synthesized in microgravity benefit from reduced convection and sedimentation effects, leading to uniform element distribution and defect-free microstructures. Containerless processing techniques, such as electromagnetic levitation, enable the formation of metallic glasses, amorphous alloys with exceptional strength, flexibility, and corrosion resistance. These materials, difficult to produce on Earth due to rapid solidification requirements, can be synthesized with greater control in microgravity. Additionally, high-entropy alloys (HEAs), which consist of five or more principal elements, have shown remarkable homogeneity and enhanced mechanical properties when processed in microgravity.

Polymers: Tailored for Space and Earth Applications

Polymers synthesized in microgravity exhibit unique properties stemming from their uniform chain alignment and reduced phase separation during polymerization. Microgravity allows for the creation of highly organized polymer nanostructures, which can lead to enhanced mechanical strength, electrical conductivity, and thermal stability. For example, conductive polymers used in space electronics have been synthesized with improved homogeneity, enabling better performance and reliability in harsh environments. Additionally, biomedical polymers, such as biocompatible hydrogels, can be tailored more precisely for targeted drug delivery and tissue engineering applications.

Challenges and Future Prospects

Despite the promise of microgravity material synthesis, significant challenges persist. The high cost of space missions, logistical limitations, and restricted experimental space on the International Space Station (ISS) pose substantial barriers. Additionally, scaling up microgravity-based manufacturing processes for terrestrial applications remains an ongoing challenge. However, emerging technologies, such as additive manufacturing, artificial intelligence (AI), and automated robotics, are steadily addressing these limitations.

Microgravity represents an unprecedented opportunity to transcend the physical limitations imposed by Earth's gravity in material synthesis. From protein crystals and quasi-crystals to metallic glasses and advanced polymers, the potential for creating groundbreaking materials is vast. As access to space continues to improve, and technological advancements drive innovation, microgravity-based material synthesis is poised to revolutionize fields ranging from pharmaceuticals and energy storage to aerospace engineering and advanced electronics. The ongoing exploration of space is not just about reaching new frontiers, it is about discovering materials that could redefine technology and human advancement for generations to come.

SIGNIFICANCE OF MICROGRAVITY RESEARCH: UNLOCKING UNIQUE PHYSICAL AND CHEMICAL PHENOMENA FOR NOVEL MATERIALS

Unique Physical and Chemical Phenomena in Microgravity

Microgravity presents a fundamentally different environment for material synthesis, where the physical and chemical processes governing crystallization, solidification, and phase separation are largely free from the gravitational forces that dominate terrestrial experiments. On Earth, gravity induces buoyancy-driven convection currents, sedimentation, and hydrostatic pressure gradients, which cause uneven mass and heat transfer during synthesis. These forces lead to imperfections such as inclusions, grain boundaries, and nonuniform microstructures in synthesized materials.

In microgravity, convection is significantly reduced, and diffusion becomes the primary mechanism for mass transport. Diffusion-driven transport allows atoms, molecules, and particles to move and arrange themselves more uniformly, leading to homogeneous compositions and precise crystal lattice formations. Without the interference of gravity-driven effects, researchers can achieve unprecedented control over processes such as nucleation, crystal growth, and phase transitions. This control is particularly advantageous in materials that require atomic-level precision, such as semiconductors, metallic glasses, and high-performance optical crystals.

Additionally, microgravity allows the study of metastable phases and exotic states of matter that cannot exist under gravitational constraints. For example, the synthesis of amorphous alloys and supercooled liquids benefits greatly from the slow and controlled cooling rates achievable in space. Likewise, the absence of sedimentation enables the uniform dispersion of nanoparticles in polymer matrices, resulting in advanced nanocomposites with exceptional mechanical and thermal properties. These phenomena not only lead to better material performance but also open new avenues for exploring material structures and behaviors that were previously inaccessible in terrestrial laboratories.

Enhanced Crystallization and Molecular Ordering

Crystallization, a cornerstone of material synthesis, is profoundly affected by gravity. On Earth, gravitational forces cause uneven temperature gradients and convective currents during crystal growth, leading to defects, inclusions, and limited crystal size. In microgravity, crystals can grow uniformly without gravitational interference, resulting in higher purity, fewer defects, and larger crystal sizes. This is particularly significant in the pharmaceutical industry, where protein crystallization plays a critical role in drug development. Crystals grown in microgravity often exhibit enhanced structural clarity, which improves the accuracy of X-ray crystallography and aids in designing more effective therapeutics.

Similarly, organic thin films and semiconductor crystals synthesized in microgravity exhibit fewer dislocations and more consistent molecular ordering. These improvements translate directly into better electronic performance, increased efficiency in energy conversion devices, and higher stability in extreme conditions.

Potential for Novel Material Properties and Structures

Microgravity creates an environment where materials can be synthesized with properties and structures that are fundamentally different from their Earth-grown counterparts. One such example is the creation of quasi-crystals, materials that exhibit ordered but non-repeating atomic structures. These exotic structures have been shown to form more perfectly in microgravity, as gravitational disruptions that might interfere with their unique atomic arrangements are minimized. Quasi-crystals possess exceptional mechanical strength, low friction coefficients, and unique thermal conductivity, making them valuable for aerospace components, coatings, and high-precision mechanical systems.

Another class of materials benefiting from microgravity is metallic glasses, amorphous alloys with no long-range atomic order. On Earth, the rapid cooling required to prevent crystallization is difficult to achieve uniformly across large samples due to convective heat transfer and sedimentation. In microgravity, however, these effects are minimized, allowing the production of bulk metallic glasses with superior mechanical properties, corrosion resistance, and magnetic characteristics.

Advancements in Nanomaterials and Composites

Microgravity also facilitates the precise synthesis of nanomaterials and nanocomposites. On Earth, nanoparticles tend to agglomerate or settle due to gravity, creating inconsistencies in composite materials. In contrast, microgravity enables uniform dispersion and alignment of nanoparticles within polymer matrices or metal alloys. This leads to nanocomposites with enhanced strength, conductivity, and thermal stability. Such materials hold promise for advanced aerospace structures, lightweight yet strong automobile components, and high-efficiency energy storage systems.

Additionally, the controlled environment of microgravity enables the creation of functionally graded materials (FGMs), where material properties transition seamlessly across regions within the same structure. FGMs are challenging to produce on Earth due to gravitational phase separation, but in microgravity, researchers can precisely control material gradients, resulting in components with tailored properties optimized for specific engineering applications.

Exploring New States of Matter and Phase Transitions

Microgravity research has also allowed scientists to explore new states of matter and phase transitions that are difficult to replicate under terrestrial conditions. For example, studies of supercooled liquids and colloidal suspensions have revealed insights into phase behavior, nucleation kinetics, and glass formation processes. These findings not only deepen our theoretical understanding of material science but also enable the development of materials with tunable mechanical, optical, and thermal properties.

Impact Across Industries

The potential applications of microgravity-synthesized materials span numerous industries, including aerospace, healthcare, renewable energy, and electronics. Advanced alloys with higher fatigue resistance, polymers with improved mechanical and chemical stability, and protein crystals with enhanced therapeutic efficacy are just a few examples of the far-reaching benefits. As humanity looks toward extended space missions, lunar bases, and Mars colonization, the ability to synthesize materials in situ using microgravity environments could revolutionize the logistics of space exploration and reduce reliance on Earth-based supply chains.

Microgravity is not just an experimental environment, it represents a transformative platform for material synthesis, unlocking pathways to novel properties, structures, and functionalities. By minimizing gravitational constraints, microgravity allows researchers to observe, understand, and manipulate physical and chemical processes at a level of precision previously unimaginable. As access to space research facilities continues to grow and technologies such as additive manufacturing and AI-driven process controls become integrated into space laboratories, the potential for groundbreaking discoveries and scalable innovations will only increase. The insights gained from microgravity research promise to redefine material science and engineering, offering solutions to terrestrial challenges while advancing our capabilities for space exploration and beyond.

OBJECTIVES: EXPLORING THE SYNTHESIS OF ORGANICS, CERAMICS, ALLOYS, POLYMERS, AND CRYSTALS IN MICROGRAVITY

The primary objective of microgravity material synthesis research is to harness the unique physical and chemical conditions of low-gravity environments to create materials with superior properties and innovative functionalities. Gravity, a fundamental force on Earth, imposes significant constraints on processes such as crystallization, solidification, polymerization, and ceramic sintering. These constraints result in imperfections, nonuniform microstructures, and phase segregation, limiting the performance and applicability of advanced materials. By exploring the synthesis of organics, ceramics, alloys, polymers, and crystals in microgravity, researchers aim to overcome these limitations, producing materials with unparalleled structural precision, enhanced properties, and entirely new architectures.

For organic materials, the goal is to enhance the crystallization of proteins, pharmaceuticals, and organic semiconductors. In microgravity, the absence of buoyancy-driven convection and sedimentation allows organic molecules to arrange themselves into highly ordered structures with fewer defects. This improved crystallinity translates directly into enhanced pharmaceutical efficacy and performance in organic electronic devices. Similarly, organic thin films synthesized in microgravity exhibit superior molecular alignment and defect-free interfaces, leading to better electronic conductivity and optical properties.

The synthesis of ceramics in microgravity focuses on achieving uniform phase distribution and minimizing defects caused by gravitational phase separation. On Earth, density gradients and uneven heat distribution during ceramic sintering often lead to structural weak points and optical impurities. Microgravity environments eliminate these constraints, allowing ceramic materials to achieve flawless microstructures and improved optical transparency. This has significant implications for advanced photonics, aerospace shielding, and high-performance thermal barrier systems.

For metallic alloys, the objective is to address challenges related to gravitational segregation, uneven solidification, and inclusions. In microgravity, alloy components remain evenly distributed during cooling and solidification, resulting in homogeneous microstructures and enhanced mechanical properties. Techniques such as electromagnetic levitation (EML) and containerless processing are used to synthesize metallic glasses, HEAs, and superalloys with exceptional strength, corrosion resistance, and thermal performance. These advanced alloys are critical for applications in aerospace, automotive, and high-performance industrial systems.

Polymers, another essential class of materials, are synthesized in microgravity to achieve uniform molecular alignment, minimize phase separation, and precisely control polymerization processes. Microgravity allows for the creation of nanostructured polymer composites with enhanced strength, electrical conductivity, and chemical resistance. Biomedical polymers synthesized in space demonstrate superior biocompatibility and mechanical stability, making them suitable for advanced drug delivery systems, implants, and scaffolds for tissue engineering.

Crystal synthesis stands as one of the most significant objectives in microgravity research. On Earth, gravitational effects cause convection currents, hydrostatic pressure gradients, and sedimentation during crystal growth, resulting in defects, impurities, and inconsistent lattice structures. In microgravity, crystals can grow undisturbed, with molecular building blocks arranging themselves in highly ordered and uniform patterns. This environment is particularly valuable for synthesizing protein crystals, perovskites, quasi-crystals, and other specialized crystal structures. Protein crystals grown in space have shown improved clarity and resolution, directly benefiting drug discovery and structural biology research. Meanwhile, perovskite crystals, essential for next-generation solar cells, exhibit fewer structural defects and enhanced light absorption properties when grown in microgravity. Quasi-crystals, with their unique non-repeating atomic patterns, have been synthesized with higher purity and stability, unlocking applications in photonics, catalysis, and aerospace materials.

Advancements in Microgravity Material and Crystal Synthesis

Over the past several decades, significant advancements have been made in synthesizing materials and crystals in microgravity environments. In organic synthesis, protein crystallization experiments aboard the ISS have demonstrated the formation of larger and more structurally consistent protein crystals. This advancement has had profound implications for pharmaceutical research, where higher-resolution protein structures are vital for drug design and therapeutic targeting. Additionally, organic semiconductor thin films produced in microgravity have exhibited reduced defect density and improved charge mobility, enhancing their performance in flexible electronics and photovoltaic applications.

In the domain of ceramics, microgravity research has enabled the development of highly uniform ceramic materials with improved optical clarity and mechanical strength. Transparent ceramics, used in laser optics and photonic devices, have been synthesized in space with reduced inclusions and structural imperfections, leading to enhanced performance. Furthermore, microgravity facilitates the creation of FGCs, where material properties transition seamlessly across different regions, a feat difficult to achieve under terrestrial gravitational constraints.

Metallic alloys synthesized in space have shown remarkable improvements in homogeneity and structural integrity. Techniques such as electromagnetic levitation have enabled the containerless processing of molten alloys, preventing contamination and ensuring controlled solidification. This has led to the production of highly uniform metallic glasses, which possess superior mechanical strength, corrosion resistance, and magnetic properties. Similarly, HEAs synthesized in microgravity exhibit homogeneous phase distribution and enhanced thermal stability, making them ideal candidates for aerospace and industrial applications.

Advancements in polymer synthesis have resulted in highly ordered nanocomposites with tailored properties. Microgravity eliminates gravitational phase separation, allowing for even dispersion of nanoparticles in polymer matrices. These nanocomposites demonstrate improved mechanical strength, thermal stability, and conductivity, with applications spanning from aerospace components to biomedical implants. Additionally, biodegradable polymers synthesized in space have shown improved structural integrity and functionality for targeted drug delivery systems.

In crystal synthesis, microgravity research has led to groundbreaking achievements in growing protein crystals, perovskites, and quasi-crystals with near-perfect atomic arrangements. Protein crystals grown aboard the ISS have contributed to more accurate drug designs, while perovskites have shown enhanced photovoltaic properties, paving the way for efficient solar energy solutions. Meanwhile, quasi-crystals synthesized in space exhibit unprecedented purity and stability, unlocking applications in high-strength coatings, photonic devices, and advanced manufacturing.

Challenges in Microgravity Material and Crystal Synthesis

Despite these advancements, significant challenges remain. The high cost of space missions and limited access to orbital laboratories constrain the scope and frequency of experiments. Additionally, the duration of microgravity exposure is often limited, especially in parabolic flights and drop towers, restricting long-term experimental studies.

Another challenge lies in the scalability of microgravity synthesis techniques for terrestrial production. While space-based experiments demonstrate proof-of-concept results, reproducing these findings on an industrial scale on Earth remains complex. Innovative engineering solutions, including hybrid manufacturing systems and AI-driven process optimization, are needed to bridge this gap.

Future Opportunities in Microgravity Research

The future of microgravity material and crystal synthesis holds immense promise. Advancements in autonomous systems, AI-driven monitoring, and 3D printing technologies will enable longer-duration and more precise experiments in space. Additive manufacturing techniques adapted for microgravity environments could facilitate on-demand production of advanced materials directly in space, reducing reliance on terrestrial supply chains for long-duration space missions.

Moreover, the commercialization of space platforms, driven by public–private partnerships, will accelerate research and development in microgravity material synthesis. Collaboration between research institutions, space agencies, and industrial partners will open pathways for scalable production, creating markets for space-synthesized materials on Earth.

The synthesis of organics, ceramics, alloys, polymers, and crystals in microgravity represents a transformative leap for material science. By overcoming gravitational constraints, researchers are unlocking unprecedented material properties and structures, paving the way for breakthroughs in aerospace, healthcare, energy, and beyond. As access to space continues to expand, the potential of microgravity research will redefine what is possible in material innovation and global technological advancement.

The Role of Gravity in the Synthesis of a "Living Molecule"

The inability to fully synthesize a "living molecule" on Earth might, in part, be influenced by the constraints imposed by gravity. While gravity is not the sole limiting factor, its effects on molecular interactions, fluid dynamics, and thermal convection can hinder the precise conditions necessary for the emergence or creation of such a molecule.

Gravity's Influence on Molecular Interactions

On Earth, gravity drives sedimentation, convection currents, and density stratification in liquid and gaseous environments. These effects can disrupt the even distribution of molecules in solution, causing them to settle or aggregate unevenly. In experiments attempting to synthesize prebiotic or self-replicating molecules,

- Uneven Distribution of Reactants: Gravity causes heavier molecules or particles to sink, while lighter ones remain suspended, potentially disrupting the uniform chemical environment needed for spontaneous molecular self-assembly.
- Diffusion Limitations: In microgravity, diffusion dominates over convection and sedimentation, allowing molecules to mix more evenly and interact in more consistent ways.
- Surface Tension Effects: On Earth, gravitational forces influence the behavior of fluids at air–liquid interfaces, potentially limiting the creation of stable protocell-like structures.

In a microgravity environment, these gravitational interferences are minimized, allowing more homogeneous mixing and longer interaction times between molecules, which may favor self-assembly or complex molecular reactions.

Thermal Convection and Energy Distribution

In Earth's gravity, thermal convection currents dominate heat transfer in liquids and gases. Warm, less-dense fluids rise, while cooler, denser fluids sink, creating dynamic circulation patterns. This process:

- Creates Temperature Gradients: Uneven heat distribution can result in localized "hot spots" or "cold zones" that might disrupt delicate molecular interactions.
- Shortens Interaction Times: Molecules may not remain in ideal thermal and chemical environments long enough to react or assemble properly.

In microgravity, convection is largely eliminated. Heat transfer occurs primarily through diffusion, leading to more stable temperature gradients and prolonged molecular interaction times. This uniformity may better support complex reactions like self-replication or catalysis.

Stability of Molecular Assemblies

In experiments aimed at synthesizing self-replicating systems or protocells, stability is critical. Gravity can destabilize delicate supramolecular structures:

- Layering and Collapse: Gravity can cause structures such as lipid bilayers (which mimic cell membranes) to collapse or deform.
- Aggregation of Molecules: Uneven gravitational pressure can force molecules to cluster in ways that prevent precise chemical arrangements.

In microgravity, these forces are negligible, potentially allowing molecular structures to form and persist in configurations more conducive to self-replication or metabolism-like activities.

Microgravity as an Ideal Laboratory for Origin-of-Life Research

Microgravity environments, such as those aboard the ISS, provide unique experimental conditions for studying prebiotic chemistry and molecular self-assembly. Scientists have already used microgravity to:

- Grow larger, defect-free crystals of proteins and other biological macromolecules.

- Study RNA folding and catalytic activity in low-gravity environments.
- Observe lipid membrane formation and stability, which are critical for protocell structures.

These findings suggest that microgravity might offer a more favorable environment for the synthesis of self-replicating and catalytic molecules, free from the disruptive effects of convection, sedimentation, and gravitational pressure.

Could Life Have Originated in a Microgravity Environment?

Some hypotheses suggest that life's building blocks may have formed in space and were delivered to Earth via meteorites or comets. In such environments,

- Gravity is negligible.
- Temperature fluctuations are minimal in certain protected environments (e.g., asteroid interiors).
- Cosmic radiation and other environmental factors could drive unique chemical reactions.

Experiments simulating space conditions have shown that amino acids, peptides, and other prebiotic molecules can form in these settings. This raises the possibility that microgravity played a pivotal role in the emergence of the earliest molecular systems that eventually led to life.

The Next Step: Space-Based Origin-of-Life Experiments

Given these insights, future experiments in space-based laboratories could focus on:

- Self-Replication Studies in Microgravity: Testing whether RNA-like molecules exhibit improved self-replication under microgravity conditions.
- Protocell Formation: Exploring whether lipid membranes or protocells form more readily and remain stable in microgravity.
- Directed Evolution in Space: Using microgravity platforms to conduct long-term evolutionary experiments on synthetic molecular systems.

Such research could provide conclusive evidence of whether gravity is, in fact, a significant barrier to synthesizing a "living molecule" on Earth.

While gravity is not the only factor preventing scientists from synthesizing a fully autonomous "living molecule," it does create physical and chemical constraints that are challenging to overcome in terrestrial laboratories. Microgravity, by eliminating sedimentation, convection, and gravitational stresses, provides an ideal environment for studying prebiotic chemistry, molecular self-assembly, and the emergence of self-replicating systems.

The continued exploration of these phenomena in space-based laboratories and microgravity environments could unlock the final pieces of the puzzle surrounding the origin of life and perhaps allow humanity to synthesize the first artificial "living molecule", a milestone that could redefine our understanding of biology, chemistry, and life itself.

CHAPTER 2

Theoretical Basis of Material Synthesis in Microgravity

Microgravity environments provide a unique setting for material synthesis, enabling scientists to study and manipulate processes that are heavily influenced by gravitational forces on Earth. The theoretical foundation for material synthesis in microgravity focuses on the fundamental effects of microgravity on material behavior, including reduced buoyancy-driven convection, absence of sedimentation, and enhanced diffusion-dominated processes. Additionally, the thermodynamics and kinetics of material synthesis under microgravity conditions reveal significant deviations from terrestrial processes, particularly in heat and mass transfer dynamics, phase separation, and crystallization behavior.

FUNDAMENTAL EFFECTS OF MICROGRAVITY ON MATERIAL BEHAVIOR

Reduced Buoyancy-Driven Convection

On Earth, temperature and density gradients in molten systems or liquid solutions create buoyancy-driven convection currents. These currents result from density differences caused by temperature variations, solute concentration gradients, or phase changes. Buoyancy forces cause lighter

DOI: 10.1201/9781003656081-2

materials to rise and denser materials to sink, disrupting uniform mixing and creating compositional inhomogeneities during material synthesis.

In microgravity, buoyancy-driven convection is almost entirely suppressed because gravitational forces are negligible. Consequently, heat and mass transfer occur primarily through diffusion, a far slower but more predictable process. This lack of convective motion allows molten metals, ceramic precursors, and polymer solutions to stabilize into more uniform distributions at the atomic or molecular level.

For example:

- Alloys: The reduction in convection prevents elemental segregation during solidification, resulting in more homogeneous alloys with fewer structural defects.
- Ceramics: Without buoyancy-driven movement, ceramic particles remain evenly dispersed in suspensions, leading to more uniform sintering and fewer structural weak points.

The absence of convective mixing also impacts the rate and mechanism of crystal growth, allowing researchers to achieve higher levels of structural order in crystallized materials.

Absence of Sedimentation

Sedimentation is the settling of heavier particles or phases under the influence of gravity, a process that is particularly problematic in suspensions, molten alloys, and polymer solutions on Earth. It leads to uneven distribution of phases, particle clustering, and stratification, causing defects, impurities, and inconsistencies in the final material structure.

In microgravity, sedimentation is virtually eliminated because gravitational forces are negligible. This allows heavier and lighter particles to remain suspended uniformly within a liquid or semi-liquid matrix for extended periods.

For example:

- Ceramics: Fine ceramic particles in suspension remain evenly distributed during the sintering process, preventing particle agglomeration and resulting in denser, more uniform ceramic structures.
- Polymers: Nanoparticles in polymer solutions are evenly dispersed, avoiding phase separation and enabling the synthesis of advanced nanocomposites with enhanced strength and thermal stability.

The absence of sedimentation also allows more controlled and precise deposition techniques, such as vapor deposition, where material layers can form without gravitational disturbances.

Enhanced Diffusion-Dominated Processes

In microgravity, diffusion becomes the dominant mode of mass and heat transfer. Diffusion is a passive process driven by concentration or thermal gradients, characterized by the random movement of atoms, molecules, or particles toward equilibrium. On Earth, diffusion is often overshadowed by convective flows, which result in uneven material mixing and unpredictable heat distribution.

In microgravity:

- Diffusion occurs uniformly across all directions, enabling consistent mass transfer.
- Temperature gradients are stabilized, preventing localized overheating or cooling during material synthesis.

For example:

- Alloys: In the solidification of molten alloys, diffusion-dominated transport ensures an even distribution of alloying elements, reducing the formation of brittle phases and compositional gradients.
- Polymers: Uniform diffusion prevents phase separation in polymer blends, leading to homogeneous polymer structures with improved mechanical and thermal properties.

This diffusion-driven mechanism also plays a crucial role in crystal nucleation and growth, where uniform diffusion enables the formation of larger, more structurally consistent crystals.

Thermodynamics and Kinetics in Zero Gravity

Microgravity also influences the thermodynamic equilibrium and kinetic pathways of material synthesis. Key factors include heat and mass transfer considerations and phase separation and crystallization behavior.

Heat and Mass Transfer Considerations

Heat transfer in microgravity occurs primarily through conduction and radiation, as convection is nearly absent. This has several implications for material synthesis:

- Slower Cooling Rates: Without gravitational convection, heat dissipates more slowly from molten materials. This slower cooling rate allows better control over crystallization and solidification processes.
- Stable Thermal Gradients: Temperature gradients are more uniform, minimizing localized thermal stresses and defects in growing crystals or solidifying alloys.

For example:

- Crystals: In protein crystallization, slower and more uniform cooling enables the formation of larger, defect-free crystals, improving their structural clarity for pharmaceutical applications.
- Alloys: Controlled heat dissipation in molten alloys reduces the formation of dendritic structures and improves phase uniformity.

Mass transfer under microgravity is primarily governed by Fickian diffusion, and the absence of convective disruptions allows for predictable mass transport. This enhanced control is particularly valuable for complex synthesis processes, such as vapor-phase deposition of thin films or the fabrication of multi-layered composites.

Phase Separation and Crystallization Behavior

Phase separation and crystallization are highly sensitive to gravitational effects. On Earth, gravitational forces can lead to phase segregation, where denser phases settle and lighter ones rise, creating compositional gradients.

In microgravity:

- Phase separation occurs primarily through thermodynamic driving forces, rather than gravitational effects.
- Crystallization pathways are determined by molecular interactions, thermal gradients, and diffusion dynamics, rather than gravitational sedimentation.

For example:

- Quasi-crystals: In microgravity, quasi-crystals with non-repeating atomic arrangements can form more perfectly, as gravitational disruptions are minimized.
- Polymers: Polymer blends achieve more uniform phase separation in microgravity, resulting in finely structured polymer matrices with tailored properties.

Additionally, nucleation processes in microgravity are more controlled, with fewer random disturbances affecting the initiation of crystal formation. This leads to larger crystal sizes, reduced dislocation densities, and enhanced structural clarity.

Implications for Material Synthesis

The interplay of reduced buoyancy-driven convection, absence of sedimentation, and enhanced diffusion processes creates an environment where thermodynamic and kinetic pathways can be precisely manipulated. This enables:

- Uniform and defect-free material structures.
- Enhanced control over nucleation and crystal growth.
- Greater reproducibility of experimental results.

Microgravity offers a unique platform to synthesize materials such as:

- Perfect single crystals for pharmaceutical and semiconductor applications.
- HEAs with enhanced thermal and mechanical properties.
- Transparent ceramics with superior optical clarity.
- Polymers and nanocomposites with tailored functionalities.

The theoretical basis for material synthesis in microgravity reveals a transformative potential for overcoming the physical and chemical limitations imposed by Earth's gravitational environment. Reduced buoyancy-driven convection, absence of sedimentation, and diffusion-dominated mass transport, combined with controlled thermodynamic and kinetic processes, provide an unparalleled platform for synthesizing materials with novel properties. As space technologies advance and access to microgravity research platforms expands, the continued exploration of these theoretical foundations will drive innovation across industries, enabling breakthroughs in fields such as aerospace, pharmaceuticals, renewable energy, and advanced electronics.

SYNTHESIS OF ORGANIC MATERIALS IN MICROGRAVITY

This section explores the significant impact of microgravity on the synthesis of organic materials, which are essential in industries like pharmaceuticals, electronics, and biomedicine. Terrestrial synthesis of these materials faces challenges due to gravitational forces that cause imperfections such as sedimentation, buoyancy-driven convection, and phase separation. Microgravity eliminates these disturbances, offering a transformative environment for more precise control over crystallization, molecular alignment, and polymerization. Key areas of focus in microgravity include organic crystals (e.g., protein crystals and pharmaceutical compounds) and polymeric materials (e.g., nanocomposites and conductive polymers). Experiments in space have demonstrated improved structural integrity, fewer defects, and enhanced functionality in materials such as protein crystals, organic semiconductors, and biodegradable polymers. These advances have profound implications for drug development, flexible electronics, and tissue engineering. Despite challenges such as high costs and scalability, innovations in automation and AI are making space-based material synthesis more accessible and reproducible. The synthesis of organic materials in microgravity is expected to lead to groundbreaking advancements in these fields, pushing the boundaries of what is achievable on Earth.

Organic materials, including proteins, pharmaceuticals, organic semiconductors, and advanced polymeric structures, play an indispensable role in industries such as pharmaceuticals, electronics, and biomedicine. However, synthesizing these materials on Earth faces persistent challenges caused by gravitational effects, including sedimentation, buoyancy-driven convection, and phase separation. These gravitational forces disrupt uniform molecular alignment, crystal growth, and polymerization processes, leading to imperfections, reduced efficiency, and diminished material performance. Microgravity offers a transformative environment for synthesizing organic materials, where gravitational disturbances are largely eliminated. This unique environment enables more precise control over crystallization, molecular ordering, and polymerization processes, creating opportunities for materials with enhanced structural integrity and defect-free compositions. Organic material synthesis in microgravity focuses primarily on two categories: organic crystals, which include protein crystals, pharmaceutical compounds, and organic optoelectronic materials, and polymeric materials, encompassing nanocomposites, conductive polymers, and biodegradable systems. These materials hold immense potential for drug development, flexible electronics, advanced sensors, and bioengineering solutions, offering breakthroughs that remain unattainable under terrestrial conditions.

In microgravity, scientists have successfully synthesized various organic compounds and crystals with significant improvements in structure and functionality. Protein crystals, essential for pharmaceutical development, enzyme studies, and biotechnology, grow with fewer defects and larger sizes in microgravity, leading to higher-resolution structural analysis for drug design. Similarly, pharmaceutical compounds requiring precise crystal lattice structures for optimal bioavailability and targeted delivery have shown marked improvements in space-based synthesis. Organic semiconductors, such as thin films used in OLEDs, organic photovoltaic (OPVs), and flexible electronic circuits, benefit from reduced defect densities and uniform molecular alignment when synthesized in microgravity. Conductive polymers, including polyaniline and poly(3-hexylthiophene) (P3HT), demonstrate improved conductivity and structural stability, enhancing their suitability for aerospace and biomedical applications. Furthermore, biodegradable polymers, such as polylactic acid (PLA) and polycaprolactone (PCL), synthesized in microgravity exhibit tailored properties for advanced tissue engineering and biocompatible medical implants. These compounds are produced using solution crystallization, thin-film deposition, emulsion polymerization, and vapor-phase polymerization, each benefiting uniquely from the absence of gravitational disturbances.

One of the most significant impacts of microgravity on organic material synthesis is observed in polymerization and crystallization processes. On Earth, density-driven convection and sedimentation cause inconsistencies in polymer chain alignment and crystal growth, often resulting in structural defects. In contrast, microgravity minimizes these effects, allowing diffusion to dominate mass transfer, leading to more uniform molecular interactions during polymerization. This ensures consistent crystal growth with fewer inclusions and more stable lattice structures. Additionally, molecular alignment in polymers synthesized in space is far more uniform, enhancing both mechanical and thermal stability. Reduced gravitational interference also minimizes defect formation in organic crystal structures, including voids, dislocations, and inconsistent molecular chain alignments, resulting in materials with enhanced optical, electronic, and mechanical properties. For instance, organic thin films synthesized in space exhibit greater consistency, improved charge carrier mobility, and fewer structural irregularities, directly translating to better performance in electronic and optoelectronic devices.

A compelling example of microgravity's impact on organic material synthesis is observed in protein crystal growth experiments conducted aboard the ISS. On Earth, protein crystals often suffer from irregularities caused by convection currents and sedimentation, limiting their structural resolution during X-ray crystallography analysis. In contrast, the microgravity environment allows protein molecules to arrange themselves into highly ordered and defect-free structures, driven purely by diffusion. In one notable study, scientists synthesized lysozyme protein crystals in microgravity and compared them to their Earth-grown counterparts. The space-grown crystals were significantly larger, more uniform, and exhibited superior diffraction quality, enabling more accurate analysis of their molecular structures. These advancements have profound implications for pharmaceutical research, as improved protein structures enhance drug design, reduce side effects, and increase therapeutic efficacy.

Another successful application of microgravity organic synthesis lies in organic semiconductor thin films. Organic semiconductors are crucial for flexible electronics, OLEDs, and photovoltaic devices, but their performance is often hampered by structural defects and uneven molecular alignment when synthesized on Earth. In microgravity, these disturbances are minimized, leading to thin films with reduced defect density, improved molecular alignment, and enhanced charge carrier mobility. Researchers aboard the ISS synthesized thin films of P3HT, observing superior uniformity, reduced impurities, and increased charge mobility compared to their Earth-grown counterparts. Similarly, experiments involving OLED thin films demonstrated fewer defects, resulting in brighter displays, higher energy efficiency, and longer operational lifespans. These improvements pave the way for more reliable and efficient electronic devices, particularly in space exploration technologies and next-generation flexible electronics.

The advancements achieved through the synthesis of organic materials in microgravity have far-reaching implications across multiple sectors. In pharmaceutical research, microgravity-grown protein crystals are accelerating drug discovery and improving the precision of therapeutic interventions. In electronics, organic semiconductors and thin films synthesized in space are setting new benchmarks for flexible, energy-efficient devices. Biodegradable polymers produced in microgravity are finding applications in tailored medical implants, drug delivery systems, and tissue scaffolding, revolutionizing biomedical engineering. However, these promising results are accompanied by significant challenges. The cost of conducting experiments in microgravity remains high, with limited access to platforms such as the ISS. Additionally, scalability remains a significant hurdle, as reproducing space-based synthesis processes in terrestrial manufacturing environments is often complex and resource-intensive. Resource availability aboard space platforms is limited, restricting large-scale experimentation and commercial production.

Despite these challenges, the opportunities presented by microgravity synthesis of organic materials continue to grow. Advances in automation, AI-driven monitoring, and additive manufacturing technologies are paving the way for more efficient and reproducible synthesis processes. With increased collaboration between space agencies, research institutions, and private industry partners, the scalability and commercialization of space-synthesized organic materials are becoming more feasible. The synthesis of organic materials in microgravity represents not just an incremental improvement in material science but a transformative breakthrough, offering pathways to materials with superior precision, reduced defect formation, and enhanced performance. As access to microgravity research platforms expands, and technological innovations drive efficiency, the synthesis of organic materials in space promises to deliver groundbreaking advancements in pharmaceuticals, flexible electronics, and biomedical engineering, fundamentally redefining what is possible in these critical fields.

SYNTHESIS OF CERAMIC MATERIALS IN MICROGRAVITY

This section discusses the synthesis of ceramic materials in microgravity, a promising approach to overcoming the limitations posed by gravity on Earth. Terrestrial ceramic synthesis faces challenges due to gravitational forces that lead to defects such as sedimentation, convection currents, and phase separation, which compromise material properties. Microgravity eliminates these effects, allowing for more uniform particle distribution, precise crystallization, and controlled sintering processes. Techniques such as physical vapor deposition (PVD) and chemical vapor deposition (CVD), isostatic sintering, and gel casting benefit from the unique environment of space, enabling the production of defect-free ceramics with superior mechanical, thermal, and optical properties. These materials have applications in high-performance optics, thermal barrier coatings (TBCs), and FGCs for aerospace and advanced military systems. Despite challenges such as the high cost and limited access to microgravity platforms, advancements in automation, additive manufacturing, and public–private partnerships hold promise for scaling up microgravity-synthesized ceramics for industrial use. These developments are poised to revolutionize sectors such as aerospace, photonics, and advanced manufacturing, contributing to future space exploration technologies.

Ceramic materials are highly valued across industries for their exceptional mechanical strength, thermal stability, resistance to wear and corrosion, and advanced optical properties. However, synthesizing defect-free ceramics on Earth remains an ongoing challenge due to the unavoidable influence of gravity on key processes such as crystallization, sintering, and phase separation. Gravitational forces cause sedimentation, where denser particles settle while lighter ones rise, leading to compositional gradients and nonuniform particle distribution. Similarly, buoyancy-driven convection in molten ceramic precursors disrupts thermal gradients and crystal growth, introducing defects such as inclusions, voids, and irregular grain structures. These gravitational effects also interfere with phase separation in multi-phase ceramic systems, causing localized aggregation of phases, which compromises structural integrity and optical clarity. Despite advancements in terrestrial techniques like hot pressing, spark plasma sintering, and sol-gel processing, gravitational constraints continue to limit the potential for synthesizing flawless ceramics.

Microgravity environments provide a transformative platform for overcoming these terrestrial limitations, offering an environment where sedimentation, convection currents, and hydrostatic pressure gradients are minimized or eliminated entirely. In space, diffusion becomes the dominant mode of mass transport, enabling particles and atoms to move uniformly and predictably during synthesis. Techniques such as vapor deposition methods and controlled sintering processes have emerged as key approaches for ceramic synthesis in microgravity. In PVD, ceramic precursors in vapor form can condense onto a substrate in a highly controlled manner, producing thin films with unparalleled surface smoothness and uniform thickness. Similarly, in CVD, microgravity eliminates uneven gas-phase transport caused by convection, allowing reactant gases to mix uniformly and produce ceramic coatings with consistent composition and minimal inclusions. These methods are particularly valuable for creating precision coatings used in optical applications, photonics, and high-temperature shielding systems.

Controlled sintering techniques in microgravity have also revolutionized ceramic fabrication. Isostatic sintering, for example, benefits from the uniform particle distribution and reduced stress gradients enabled by microgravity, producing dense and defect-free ceramic structures. Likewise, gel casting in microgravity eliminates phase separation caused by density differences, ensuring a homogeneous particle distribution in ceramic green bodies. This significantly reduces warping, cracking, and porosity during drying and sintering, leading to structurally superior ceramics. Moreover, the emergence of additive manufacturing (3D printing) technologies in space has allowed for the precise layer-by-layer deposition of ceramic precursors. In microgravity, complex geometries and FGCs can be synthesized with seamless transitions between regions of varying material properties.

The applications of microgravity-synthesized ceramics span multiple high-value sectors. In the field of high-performance optical materials, transparent ceramics such as aluminum oxynitride (ALON) and yttrium aluminum garnet (YAG) have been produced with fewer inclusions and superior optical clarity. These materials are integral to advanced imaging systems, high-precision lenses, and laser components. Similarly, ceramic optical fibers synthesized in space demonstrate reduced structural defects, resulting in improved signal transmission and data transfer capabilities, critical for satellite communications and remote sensing technologies. Another key application lies in TBCs, where ceramic materials synthesized in microgravity exhibit superior thermal resistance and mechanical adhesion. These coatings are essential for protecting turbine blades in aerospace engines and rocket propulsion systems, as they can withstand extreme thermal cycling and oxidative environments.

Furthermore, microgravity research has enabled the development of FGCs, where material properties change seamlessly across regions. These materials are essential for creating components that need to withstand varying thermal and mechanical loads, such as aerospace shielding and structural elements for spacecraft. Additionally, ceramic-matrix composites (CMCs) synthesized in space have shown enhanced structural stability, reduced porosity, and improved resistance to extreme conditions. These composites are lightweight yet robust, making them ideal for applications in spacecraft, hypersonic vehicles, and advanced military armor systems.

Despite these significant advancements, several challenges remain. Access to microgravity research platforms such as the ISS is costly and logistically complex, limiting the scale and frequency of experiments. The experimental timeframe is often constrained by mission durations, posing limitations on long-term studies. Additionally, while space-based synthesis can demonstrate proof-of-concept results, translating these findings into large-scale terrestrial manufacturing processes remains a significant hurdle. Scalability challenges arise because microgravity eliminates gravitational effects that are reintroduced during Earth-based industrial replication. However, the integration of AI-driven automation systems and additive manufacturing technologies shows promise in addressing these scalability concerns.

The future of ceramic synthesis in microgravity is bright, with significant opportunities emerging from advancements in autonomous in-space manufacturing, real-time process monitoring, and AI-optimized synthesis protocols. These innovations could enable longer-duration experiments, in situ ceramic fabrication in space, and more reliable scalability for Earth-based industrial applications. Additionally, growing public–private partnerships are expected to play a key role in accelerating research, lowering costs, and fostering commercial applications of microgravity-synthesized ceramics.

These advancements are not only revolutionizing aerospace engineering, photonics, and high-temperature applications but are also paving the way for new discoveries in advanced manufacturing and space exploration. As microgravity platforms become more accessible and experimental technologies continue to evolve, the role of space-based ceramic synthesis in shaping future technologies will become increasingly significant.

SYNTHESIS OF METALLIC ALLOYS IN MICROGRAVITY

This section explores the synthesis of metallic alloys in microgravity, highlighting the benefits of space-based alloy production in overcoming challenges posed by gravity on Earth. Gravitational forces during solidification processes, such as sedimentation, buoyancy-induced convection, and density-driven segregation, lead to structural defects like uneven compositions, dendritic formations, and voids. Microgravity eliminates these gravitational effects, enabling uniform solidification and precise control over alloy microstructures. Techniques such as electromagnetic levitation and containerless processing, including acoustic and electrostatic levitation, facilitate defect-free alloy synthesis, particularly for titanium-aluminum (Ti-Al) and nickel-based (Ni-based) superalloys. These alloys exhibit improved mechanical properties, such as enhanced strength, thermal fatigue resistance, and fine grain structures, making them ideal for aerospace and high-temperature applications. Microgravity also enables the formation of metallic glasses, which offer exceptional strength and corrosion resistance for use in high-durability components. While microgravity alloy synthesis holds significant promise for industries like aerospace, automotive, and precision instrumentation, challenges remain in scaling up production and translating space-based results to Earth-based manufacturing. Future advancements in AI-driven automation, real-time data monitoring, and commercial spaceflight are expected to enhance alloy synthesis capabilities and lower costs, paving the way for large-scale, space-based alloy production.

The synthesis of metallic alloys in microgravity represents a significant advancement in material science, offering an opportunity to overcome the inherent challenges posed by gravitational forces during solidification processes. Alloys, which are engineered combinations of two or more metallic elements, are essential materials in aerospace, automotive, and advanced engineering applications due to their strength, durability, and versatility. However, on Earth, the process of alloy solidification is heavily influenced by gravity, leading to a range of structural and compositional defects. Gravitational forces drive buoyancy-induced convection, sedimentation, and density-driven segregation, all of which disrupt the uniform distribution of alloying elements and influence the growth of crystal structures. As molten alloys cool and solidify under gravity, heavier alloying elements may settle while lighter ones rise, causing macrosegregation, a phenomenon where different regions of the alloy exhibit varying compositions. Additionally, uneven thermal gradients and convective flows result in inconsistent crystal growth, introducing defects such as dendritic structures, inclusions, and voids. These imperfections compromise the mechanical and thermal properties of the alloy, limiting its performance in critical applications.

In a microgravity environment, these gravitational effects are significantly reduced or eliminated, creating ideal conditions for uniform solidification and precise control over alloy microstructures. One of the most effective techniques employed in space-based alloy synthesis is EML. In this method, molten metal droplets are suspended and manipulated using electromagnetic fields, ensuring that the material remains in a stable, contactless state throughout the solidification process. The absence of physical contact with container walls eliminates the risk of contamination and minimizes heterogeneous nucleation, where impurities act as crystallization sites. Electromagnetic levitation also allows for precise control of temperature and solidification rates, enabling the formation of highly uniform microstructures with fewer defects. Another widely used approach is containerless processing, where molten alloys are suspended in space without the need for a physical container. Techniques such as acoustic levitation and electrostatic levitation further enhance the purity and structural consistency of the alloy by preventing interactions with container walls and minimizing external thermal gradients. These containerless methods enable the production of bulk metallic glasses, which are highly desirable due to their amorphous structures, exceptional mechanical strength, and corrosion resistance.

Space-based alloy synthesis experiments have yielded several groundbreaking results, particularly with Ti-Al and Ni-based superalloys. Ti-Al alloys are widely used in aerospace applications due to their high strength-to-weight ratios and resistance to oxidation at elevated temperatures. Under terrestrial conditions, the solidification of Ti-Al alloys is prone to gravitational segregation, resulting in uneven phase distributions and brittle intermetallic phases. In microgravity, however, the absence of sedimentation and buoyancy-induced convection allows Ti-Al alloys to solidify with enhanced phase uniformity, leading to improved mechanical properties and resistance to thermal fatigue. Similarly, Ni-based superalloys, which are essential for high-temperature turbine blades and jet engine components, have shown remarkable improvements when synthesized in microgravity. Space-grown Ni-based superalloys exhibit finer grain structures, fewer dendritic formations, and homogeneous element distribution, resulting in superior thermal and mechanical performance.

Another noteworthy achievement in microgravity alloy synthesis is the formation of amorphous metals, also known as metallic glasses. On Earth, metallic glass formation requires extremely rapid cooling rates to prevent crystallization, a process that is difficult to achieve uniformly across large samples due to thermal gradients and convection currents. In microgravity, these effects are minimized, allowing molten metal alloys to cool uniformly and remain in an amorphous state. Metallic glasses synthesized in space exhibit unmatched strength, elasticity, and resistance to corrosion, making them ideal for applications requiring high durability and precision, such as gears, springs, and wear-resistant coatings.

The applications of space-synthesized metallic alloys extend across multiple high-performance sectors, with the aerospace industry being one of the primary beneficiaries. Lightweight alloys with improved mechanical properties, such as Ti-Al superalloys, are critical for reducing the weight of aircraft and spacecraft components without compromising structural integrity. Ni-based superalloys, with their superior thermal resistance, are essential for the development of more efficient and reliable turbine blades in jet engines. The high thermal and mechanical stability of these alloys enables them to withstand extreme operational environments, including rapid temperature fluctuations and mechanical stresses encountered during high-speed flight and re-entry phases.

Beyond aerospace, space-synthesized alloys hold promise for the development of high-strength, lightweight structural materials used in automotive manufacturing, precision instrumentation, and energy systems. The enhanced strength-to-weight ratios, uniform microstructures, and reduced defect densities achieved in microgravity make these alloys ideal for applications where weight savings and mechanical performance are critical. Additionally, metallic glasses produced in space are finding applications in precision components such as microgears, actuators, and sensor systems, where their amorphous structures provide exceptional mechanical stability and resistance to wear.

Despite the groundbreaking advancements achieved in microgravity alloy synthesis, several challenges remain. The high cost and logistical complexity of conducting experiments in space continue to limit the frequency and scale of these projects. Additionally, scaling up microgravity-based alloy synthesis for terrestrial mass production remains a significant challenge, as gravitational forces are reintroduced in Earth-based environments. Researchers are exploring hybrid manufacturing approaches that combine insights from space-based experiments with advanced terrestrial processing techniques to replicate microgravity conditions on a smaller scale.

Looking ahead, advancements in AI-driven automation, real-time data monitoring, and additive manufacturing techniques are expected to play a key role in enhancing alloy synthesis capabilities in space. Autonomous systems capable of operating in microgravity could enable long-duration experiments, ensuring precise control over solidification parameters and minimizing resource dependency on human intervention. As commercial spaceflight becomes more accessible and private-sector involvement in space manufacturing increases, the development of space-based alloy synthesis facilities could become a reality, driving down costs and scaling up production.

SYNTHESIS OF POLYMERS IN MICROGRAVITY

The synthesis of polymers in microgravity presents a transformative approach to overcoming the limitations imposed by Earth's gravitational forces during polymerization. Gravity-induced effects such as phase separation, sedimentation, and uneven filler dispersion compromise the structural and functional integrity of Earth-grown polymers. In contrast, the microgravity environment facilitates uniform diffusion, consistent thermal distribution, and homogeneous molecular alignment, enabling the production of polymers with superior mechanical, thermal, and chemical properties. This chapter explores the advantages of emulsion and controlled radical polymerization (CRP) techniques under microgravity conditions, highlighting their ability to yield polymers with precise architectures and enhanced performance characteristics. Applications span aerospace, biomedicine, and electronics, with notable improvements in tensile strength, thermal stability, and conductivity. Despite current logistical challenges and scalability issues, emerging technologies in automation, remote sensing, and additive manufacturing hold promise for expanding microgravity polymer synthesis and translating its benefits to Earth-based production systems.

Polymers are essential materials in modern technology, spanning applications in aerospace, biomedicine, electronics, and industrial manufacturing. Their versatility arises from their unique mechanical, thermal, and chemical properties, which can be tailored through precise control of polymerization processes and molecular architecture. However, the synthesis of polymers on Earth is inherently influenced by gravity, which introduces density gradients, phase separation, and sedimentation effects during polymerization. These gravitational forces disrupt molecular alignment, result in phase inhomogeneities, and often cause defects in the final polymer structure. For example, polymer blends and composites frequently suffer from uneven dispersion of fillers or nanoparticles, leading to inconsistent mechanical properties and reduced performance. Additionally, gravitational effects during phase transitions introduce voids and structural irregularities, which degrade the final product's mechanical stability and thermal behavior.

Microgravity offers a unique environment where these disruptive gravitational effects are minimized or eliminated, allowing for more precise control over polymerization processes. In microgravity, diffusion becomes the dominant mechanism of mass transport, enabling uniform mixing of reactants and consistent heat dissipation. This environment creates the conditions necessary for the development of polymers with highly homogeneous structures, tailored thermal and mechanical properties, and enhanced performance characteristics. Two primary techniques have been employed to leverage these benefits: emulsion polymerization and CRP.

Emulsion polymerization is a process in which monomers are dispersed in water with the help of surfactants, forming small droplets that act as individual polymerization reactors. On Earth, gravity causes density gradients in these droplets, resulting in uneven polymer particle sizes and inconsistent polymer chain lengths. In microgravity, these density-driven effects are largely absent, allowing droplets to remain uniformly dispersed throughout the reaction medium. This uniform dispersion results in more consistent polymer particle sizes, improved molecular weight distributions, and enhanced polymer purity. For instance, emulsions in microgravity have been shown to produce nanoparticles and microgels with significantly fewer defects, opening possibilities for highly uniform nanostructured polymer systems.

CRP is another technique that benefits greatly from microgravity conditions. CRP enables precise control over polymer chain growth, molecular weight, and architecture. On Earth, gravitational forces often create phase separation and nonuniform heat distribution, limiting the degree of control over chain propagation and termination. In microgravity, these effects are minimized, allowing polymer chains to grow in a highly controlled manner. This leads to polymers with predictable chain lengths, consistent branch architectures, and enhanced end-group functionalities. These improvements are particularly valuable for synthesizing block copolymers, star-shaped polymers, and hyperbranched polymers with highly specific properties for advanced applications.

The polymers synthesized in microgravity exhibit several unique properties that set them apart from their Earth-grown counterparts. One of the most significant benefits is their enhanced structural homogeneity. The elimination of gravitational effects during polymerization ensures that polymer chains grow uniformly and consistently, minimizing chain entanglements, voids, and phase separation. This uniformity translates directly into improved mechanical properties, such as higher tensile strength, increased elasticity, and enhanced resistance to stress fractures. Similarly, the thermal stability of microgravity-synthesized polymers is superior due to the even distribution of thermal gradients during polymerization. These polymers also demonstrate tailored mechanical properties, which can be customized for specific applications, including flexibility for wearable electronics, stiffness for structural components, or biocompatibility for medical devices.

The superior properties of polymers synthesized in microgravity have paved the way for applications in medicine, aerospace, and electronics, where performance and reliability are critical. In biomedicine, space-grown polymers are being explored for use in biomedical implants, drug delivery systems, and tissue scaffolding. For example, biocompatible polymers such as PLA and PCL synthesized in space exhibit highly uniform porosity and controlled degradation rates, making them ideal for implants and tissue engineering scaffolds. Their enhanced structural homogeneity reduces the risk of implant failure, while controlled biodegradability ensures consistent performance over time. Similarly, polymer hydrogels synthesized in microgravity offer superior swelling properties and drug encapsulation efficiency, revolutionizing targeted drug delivery systems.

In the aerospace industry, lightweight yet durable polymers are critical for reducing weight and improving efficiency in spacecraft and satellite systems. Space-synthesized polymers demonstrate exceptional resistance to radiation, extreme temperatures, and mechanical stress. These properties make them ideal for structural components, thermal insulation, and protective coatings on spacecraft. Additionally, polymers with highly tailored mechanical flexibility and conductivity are being explored for use in deployable structures and adaptive aerospace materials.

In electronics, space-synthesized polymers are finding applications in conductive polymers for space electronics and advanced flexible circuits. Conductive polymers such as polyaniline and P3HT synthesized in microgravity exhibit superior electrical conductivity and reduced structural defects, leading to improved charge carrier mobility. These properties are particularly valuable for the development of lightweight, flexible electronic components used in space-based applications, including sensors, photovoltaic arrays, and radiation-resistant circuitry. Additionally, microgravity allows for the synthesis of polymers with tunable dielectric properties, opening avenues for high-performance capacitors and energy storage devices.

Despite these advancements, challenges remain in the synthesis of polymers in microgravity. Logistical and financial constraints associated with conducting polymerization experiments aboard platforms like the ISS limit the frequency and scale of research efforts. Additionally, the scalability of space-based polymerization processes for terrestrial production remains a significant hurdle. Translating the precise control achieved in microgravity environments to large-scale, Earth-based manufacturing processes often introduce gravitational disturbances, reducing reproducibility.

Looking forward, advancements in AI-driven automation, remote sensing technologies, and additive manufacturing techniques promise to address these limitations. Automation systems capable of real-time monitoring and control during polymerization processes could enable longer-duration experiments with minimal human intervention. Similarly, additive manufacturing techniques adapted for microgravity could facilitate the in-situ fabrication of polymer-based components directly aboard space missions.

CHAPTER 3

Experimental Facilities and Platforms for Microgravity Research

Microgravity research has been revolutionized by the development of specialized platforms that enable the study of material synthesis under reduced gravity conditions. From the long-duration capabilities of the International Space Station (ISS) to the brief, controlled environments of parabolic flights and drop towers, each platform offers unique advantages and limitations for investigating processes such as polymerization, crystallization, alloy solidification, and ceramic sintering. The ISS serves as a premier orbital laboratory with advanced instrumentation for high-precision, long-term experiments, while parabolic flights and drop towers enable rapid prototyping and short-duration studies. Autonomous satellites and space-based laboratories further extend microgravity research by providing intermediate-duration, cost-effective experimental opportunities. These platforms have facilitated critical advancements across aerospace, biomedical, and materials science domains, but challenges remain, including high launch costs, limited real-time monitoring, and logistical constraints. Emerging solutions, such as AI-driven automation, additive manufacturing, and improved communication infrastructure, are poised to enhance platform capability and acces-

DOI: 10.1201/9781003656081-3

sibility. Collectively, these microgravity research platforms are expanding the frontiers of materials science and engineering, enabling discoveries with profound implications for both space exploration and terrestrial technologies.

The advancement of microgravity research owes much to the development of specialized facilities and platforms designed to enable material synthesis experiments in low-gravity environments. These platforms, ranging from orbital laboratories like the ISS to transient microgravity facilities such as parabolic flights and drop towers, offer scientists unique opportunities to study fundamental material behavior in ways not possible on Earth. Each facility has distinct capabilities, durations, and limitations, contributing to our growing understanding of how microgravity influences processes like crystallization, polymerization, alloy solidification, and ceramic sintering. While these platforms have pushed the boundaries of material science, they also come with technological challenges that must be addressed to fully realize their potential.

The ISS stands as the flagship platform for long-duration microgravity experiments. Orbiting approximately 400 kilometers above Earth, the ISS provides an environment where gravitational effects are reduced to nearly zero (10^{-6} g). The station is equipped with sophisticated research modules such as NASA's Destiny Laboratory, ESA's Columbus Laboratory, and JAXA's Kibo Laboratory, each housing specialized equipment for material synthesis and analysis. Instruments like the Electromagnetic Levitator and the Materials Science Laboratory (MSL) allow scientists to conduct precise experiments on alloy solidification, ceramic sintering, and crystal growth. For example, the Electromagnetic Levitator enables containerless processing of metallic alloys, preventing contamination and facilitating highly controlled solidification experiments. The ISS also hosts advanced protein crystallization facilities, enabling pharmaceutical companies to grow protein crystals with unmatched structural clarity. Furthermore, its long-duration operational capability allows researchers to study slow material processes that cannot be replicated in shorter-duration microgravity platforms. However, conducting experiments aboard the ISS is not without challenges. Limited cargo space, the high cost of launching experimental payloads, and scheduling constraints due to mission priorities often restrict the scope and scale of research activities.

In contrast to the extended microgravity conditions aboard the ISS, parabolic flights and drop towers provide short-duration, intermittent access to microgravity environments. Parabolic flights, typically conducted using specially modified aircraft such as the Airbus A310 Zero-G, create periods of weightlessness lasting approximately 20–25 seconds per parabola. During each maneuver, the aircraft climbs steeply before entering a free-fall phase, simulating near-weightless conditions. These flights are particularly useful for preliminary studies, prototype testing, and validating experimental setups before deployment on the ISS. Drop towers, on the other hand, offer an even shorter but highly controlled microgravity environment, typically lasting 2–10 seconds. Facilities like the Bremen Drop Tower in Germany enable researchers to conduct quick, repeatable experiments with exceptional precision. These platforms are ideal for studying rapid material processes, such as the initial stages of nucleation during crystallization or transient phase transitions in polymerization. However, their limited microgravity duration restricts their applicability to experiments requiring longer observation periods.

Beyond the ISS and short-duration facilities, space-based research laboratories and satellites provide an intermediate solution for conducting microgravity experiments over weeks or months. Autonomous satellites, such as CubeSats and NanoSats, are increasingly being used for specialized microgravity experiments. These compact satellites are deployed into orbit with pre-programmed experimental protocols and often carry automated systems for data collection and analysis. For example, NanoRacks' CubeLabs has been utilized to conduct crystallization and thin-film deposition experiments in low-Earth orbit (LEO). Dedicated research platforms, such as China's Tiangong Space Station, are also emerging as important contributors to microgravity research, offering advanced facilities and longer operational durations. These platforms offer a balance between the cost-effectiveness of parabolic flights and the extended operational capacity of the ISS. However, challenges such as limited power supply, communication latency, and restricted real-time monitoring capabilities continue to pose barriers to fully utilizing these platforms.

While microgravity research platforms have advanced significantly, several technological challenges and innovations continue to shape their development and application. Automation and remote monitoring have become essential for maximizing the efficiency of experiments in space. Advanced AI-driven monitoring systems are being developed to enable autonomous adjustments and real-time analysis, reducing dependence on astronauts for routine interventions. Similarly, miniaturization of laboratory equipment is a growing trend, allowing more compact, lightweight experimental setups to be launched into space. Innovations in additive manufacturing technologies, such as 3D printing in space, are also emerging as transformative tools for microgravity research. These technologies allow for the creation of custom experimental tools, components, and materials directly aboard space platforms, reducing reliance on pre-launch cargo.

Despite these advancements, logistical and financial barriers remain significant hurdles. Launching payloads into space remains prohibitively expensive, and experimental slots aboard platforms like the ISS are highly competitive. Additionally, data transmission delays and limited bandwidth can restrict real-time monitoring and control of space-based experiments. To address these issues, new technologies, including low-latency communication systems and cloud-based data platforms, are being developed to streamline data collection and analysis. Furthermore, the growing role of public–private partnerships is helping to democratize access to microgravity platforms. Companies like SpaceX, Blue Origin, and Northrop Grumman are increasingly offering commercial launch services and microgravity experiment opportunities, driving down costs and expanding accessibility for researchers.

The ISS, parabolic flights, drop towers, and autonomous satellites each play a unique role in advancing microgravity research. While the ISS offers long-duration experimental opportunities, parabolic flights and drop towers provide short but highly controlled microgravity windows ideal for rapid testing. Space-based autonomous laboratories bridge the gap, offering cost-effective and moderately extended access to low-gravity environments. Each platform contributes invaluable insights into the behavior of materials under microgravity conditions, enabling advancements in fields ranging from pharmaceutical crystallization to advanced alloy and polymer synthesis. As technologies such as AI-driven automation, additive manufacturing, and low-latency communication continue to mature, the efficiency and accessibility of microgravity research will only improve. These platforms are not merely experimental tools, they represent humanity's collective ambition to push the boundaries of material science and engineering, unlocking innovations that will shape the technologies of tomorrow.

CHALLENGES AND LIMITATIONS IN SPACE MATERIAL SYNTHESIS

Space-based material synthesis offers unprecedented opportunities to engineer materials with enhanced structural, thermal, and functional properties by leveraging the unique conditions of microgravity. However, its transformative potential is tempered by significant challenges, including high costs, logistical constraints, limited experimental access aboard space platforms, and difficulties in scaling microgravity-optimized processes for terrestrial applications. Safety and environmental considerations further complicate implementation. Despite these hurdles, emerging technologies, such as AI-driven automation, additive manufacturing, and real-time process monitoring, are poised to revolutionize space manufacturing. Advances in autonomous systems, in-situ resource utilization (ISRU), and space-adapted 3D printing are enabling the vision of fully automated, orbiting material synthesis laboratories and space-based factories. These innovations support the scalable production of high-performance alloys, defect-free crystals, transparent ceramics, and advanced polymers, with far-reaching implications for aerospace, pharmaceuticals, electronics, and sustainable manufacturing. As launch costs decrease and public–private partnerships proliferate, the commercialization of space-grown materials is becoming increasingly viable. The future of space material synthesis lies not only in scientific discovery but in redefining the boundaries of industrial capability, innovation, and human achievement both in space and on Earth.

Despite the transformative potential of space-based material synthesis, significant challenges and limitations persist, hindering the widespread adoption and scalability of these technologies. These obstacles arise from the cost and logistical constraints of space missions, the limited availability of experimental time and resources aboard space platforms, the difficulty of scaling microgravity-optimized processes for terrestrial production, and safety and environmental concerns associated with space-based experimentation. Addressing these challenges will require technological innovation, strategic collaboration, and policy support to fully unlock the potential of space material synthesis.

The cost and logistics of space missions remain one of the most formidable barriers to large-scale material synthesis in microgravity environments. Launching payloads into space is extraordinarily expensive, with costs running into tens of thousands of dollars per kilogram. These expenses include not only transportation but also the design, testing, and validation of experimental hardware to meet the rigorous safety and operational standards required for space environments. Experiments must be meticulously engineered to withstand the mechanical stresses of launch, the vacuum of space, and the temperature fluctuations of LEO. Moreover, logistical constraints extend beyond financial considerations. Scheduling access to microgravity platforms, particularly the ISS, is subject to intense competition and limited availability. Research missions often face delays due to competing priorities, hardware malfunctions, or shifting mission goals. The dependency on specialized launch providers and the finite cargo space aboard resupply missions further complicates planning and execution. While private aerospace companies such as SpaceX, Blue Origin, and Northrop Grumman have made significant strides in reducing launch costs and increasing access to space, the logistical hurdles remain substantial for many researchers and organizations.

Another major challenge is the limited experimental time and resources available aboard microgravity platforms. On the ISS, for instance, crew members have strict schedules, balancing scientific research with station maintenance, life-support operations, and mission-critical tasks. As a result, experiments often face time constraints, limiting the ability to monitor long-duration material synthesis processes or to conduct iterative testing. Additionally, physical space aboard the ISS is at a premium, with laboratory modules packed with diverse scientific equipment, leaving little room for large-scale material synthesis apparatus. Power availability, cooling systems, and experimental consumables are also limited, further constraining the scope of research activities. Similarly, short-duration platforms like parabolic flights and drop towers offer only fleeting windows of microgravity, typically ranging from seconds to minutes, which are insufficient for many material synthesis processes requiring extended reaction times. While autonomous platforms such as CubeSats and NanoSats offer a more cost-effective alternative, they lack the real-time human oversight and sophisticated equipment available aboard crewed space stations.

Beyond the operational challenges in space, scaling up microgravity-optimized processes for terrestrial applications remains a significant limitation. Many of the material properties and structural advantages observed in microgravity-synthesized materials, such as enhanced crystallinity, uniform phase distribution, and reduced defects, are directly tied to the unique physical conditions of space. Reproducing these conditions on Earth is exceedingly difficult, if not impossible, with current technologies. For example, in alloy solidification, the absence of buoyancy-driven convection and sedimentation in space plays a critical role in achieving homogenous microstructures. Simulating these effects terrestrially would require energy-intensive setups, such as advanced EML systems or controlled-atmosphere furnaces, which are expensive and complex to maintain. Similarly, crystal growth experiments in space have shown exceptional results due to diffusion-dominated mass transport, a phenomenon challenging to replicate in Earth-based laboratories. This scalability gap limits the potential for commercial adoption of space-synthesized materials, as the cost and complexity of reproducing microgravity conditions on Earth offset the material performance gains. Researchers are actively exploring hybrid solutions, such as using machine learning algorithms and real-time process monitoring, to bridge this scalability gap, but widespread industrial adoption remains elusive.

Safety and environmental concerns also pose critical challenges to space material synthesis. Operating in space inherently carries risks, both to crew members aboard platforms like the ISS and to the broader environment. Experiments involving volatile chemical precursors, high-temperature processes, or reactive gases must be carefully contained and monitored to prevent leaks, fires, or other hazardous incidents in the confined environment of a space station. The absence of natural convection in microgravity also increases the risk of hazardous material accumulation, as toxic vapors or particles do not naturally disperse. Furthermore, the disposal of experimental by-products and chemical waste generated during space-based material synthesis is a growing concern. Space missions must adhere to strict environmental regulations to prevent the uncontrolled release of contaminants into the spacecraft environment or the surrounding space. On a broader scale, the environmental footprint of space activities, including rocket launches and orbital debris, is an increasing global concern. The carbon emissions associated with launch vehicles and the accumulation of non-functional satellites and debris in orbit highlight the need for sustainable practices in space-based research and manufacturing.

Addressing these challenges requires a multi-faceted approach. Technological innovations such as AI-driven process optimization, automated monitoring systems, and miniaturized laboratory modules can maximize the efficiency of microgravity experiments and reduce dependence on crew involvement. Meanwhile, advancements in ISRU and additive manufacturing technologies hold promise for reducing logistical constraints by enabling on-site fabrication of experimental tools and materials aboard space platforms. Collaborative partnerships between space agencies, research institutions, and private sector stakeholders will also be critical in distributing costs, sharing expertise, and accelerating research timelines. Additionally, regulatory frameworks must evolve to ensure safe and environmentally responsible practices in space-based material synthesis.

While the cost, logistical barriers, limited resources, scalability challenges, and safety concerns pose significant hurdles to space material synthesis, they are not insurmountable. Continued advancements in automation, AI, additive manufacturing, and space logistics offer promising pathways for overcoming these challenges. Furthermore, increased investment, both public and private, in space infrastructure and microgravity research platforms will play a crucial role in democratizing access to space-based material synthesis opportunities. As these barriers are addressed, the full potential of space material synthesis will be realized, offering groundbreaking advancements in aerospace engineering, pharmaceuticals, flexible electronics, and sustainable manufacturing technologies. The future of space-based material synthesis is not just an academic pursuit but a critical enabler of the next generation of technological innovation, both on Earth and beyond.

Future Prospects and Emerging Technologies in Space Material Synthesis

The future of space material synthesis is deeply intertwined with advancements in automation, AI, additive manufacturing, space-based factories, and commercialization strategies. These technologies and long-term visions aim to address the challenges of cost, scalability, and operational limitations while unlocking new opportunities for innovation, sustainability, and economic growth. As space research platforms evolve, these emerging technologies will play a pivotal role in transforming space-based material synthesis from isolated experiments into fully integrated industrial processes.

Automation and AI in Space Manufacturing are at the forefront of this transformation, promising to optimize complex experimental procedures, reduce reliance on human intervention, and maximize resource efficiency aboard space platforms. Space missions often face constraints in terms of time, available crew members, and real-time monitoring capabilities. AI-driven systems equipped with machine learning algorithms can autonomously monitor and adjust experimental parameters in real time, ensuring optimal conditions for processes like alloy solidification, polymerization, and crystal growth. For instance, AI can predict the behavior of molten alloys in EML systems, dynamically fine-tuning temperature and cooling rates to minimize defects. Automation also reduces human error and enables round-the-clock monitoring of long-duration experiments. On platforms like the ISS, robotic systems could perform complex sample handling, processing, and data collection tasks, freeing astronauts to focus on high-level mission priorities. Furthermore, AI can analyze vast datasets generated from microgravity experiments, identifying patterns and optimizing process parameters for future iterations. As AI systems continue to improve, they will be instrumental in enabling fully autonomous material synthesis laboratories in space, capable of conducting experiments, analyzing results, and even initiating corrective measures without direct human oversight.

In parallel with automation and AI, additive manufacturing and 3D printing in microgravity are emerging as revolutionary technologies that could redefine how materials are synthesized and utilized in space. Traditional manufacturing techniques rely on gravitational forces for material deposition, phase separation, and structural stability. In microgravity, these processes must be reimagined. 3D printing technologies adapted for microgravity environments can overcome these constraints by building complex material structures layer by layer, directly aboard space platforms. This capability allows for on-demand production of advanced components, tools, and even experimental apparatuses without requiring costly resupply missions from Earth. For example, CMCs and metallic alloys could be synthesized using space-based 3D printing systems, allowing for precise control over material composition and structural integrity. Similarly, biopolymer scaffolds for medical implants could be printed with highly controlled porosity and tailored mechanical properties. Additive manufacturing in space also reduces material waste, as it can utilize ISRU to recycle and repurpose available materials. As 3D printing technologies continue to advance, they will play a vital role in creating self-sustaining ecosystems for material synthesis in space, minimizing reliance on terrestrial supply chains.

The long-term vision for space-based factories represents the next frontier in material synthesis. Current research platforms, such as the ISS and parabolic flights, are primarily geared toward scientific exploration and proof-of-concept experiments. However, the vision of dedicated space-based manufacturing hubs is gaining momentum. These future facilities could be fully automated, AI-driven manufacturing plants orbiting Earth or stationed on the Moon and Mars, optimized for continuous production of advanced materials in microgravity. Space factories would not only focus on producing high-performance alloys, transparent ceramics, and tailored polymers but also on scaling these processes for commercial viability. For example, metallic glasses synthesized in zero gravity could be produced in bulk and transported back to Earth for use in industries requiring ultra-durable and lightweight materials. Lunar regolith, rich in silicates and metals, could serve as raw material for space-based ceramic and alloy production. Such facilities might also integrate modular robotic systems and automated quality control mechanisms to ensure scalability, consistency, and precision in material synthesis. In the long term, these space-based factories could become nodes in an interplanetary supply chain, supporting extended human presence in space and enabling construction projects on the Moon, Mars, and beyond.

The potential for commercialization and market impact of space material synthesis is immense. While space-based material research has largely been limited to government-funded agencies and academic institutions, the growing involvement of private-sector players such as SpaceX, Blue Origin, and Northrop Grumman is shifting the landscape. These companies are driving down the cost of space access, making commercial material synthesis experiments more economically viable. Markets for space-grown pharmaceuticals, high-purity protein crystals, defect-free semiconductors, and advanced alloys are projected to grow significantly in the coming decades. Pharmaceutical companies, for instance, are already exploring space-based protein crystallization to develop more effective drugs with fewer side effects. The electronics industry is interested in defect-free organic semiconductors synthesized in microgravity for use in flexible electronics, OLED displays, and advanced photovoltaic systems. Similarly, aerospace and automotive industries are eyeing space-produced metallic glasses and HEAs for lightweight, ultra-durable components.

However, commercial scalability will depend on overcoming current logistical and cost-related barriers. As the cost of launches continues to decrease and reusable rocket technologies become more common, space-based manufacturing may reach a tipping point where it becomes financially feasible for mass production. Public–private partnerships will play a critical role in this transition, with governments and private entities working together to share costs, risks, and expertise. Investment in ground-based simulation technologies, AI-enhanced process optimization, and

Earth-based scalability solutions will be equally crucial for ensuring that space-derived technologies can be adapted for terrestrial applications.

The future of space material synthesis is filled with promise, driven by advancements in automation, AI, additive manufacturing, and the establishment of space-based factories. These emerging technologies are set to transform how materials are synthesized, tested, and applied, not only in space but also on Earth. As the boundaries between scientific research and commercial applications continue to blur, space material synthesis is poised to become a multi-billion-dollar industry with profound implications for aerospace engineering, pharmaceuticals, renewable energy, electronics, and advanced manufacturing. The realization of these prospects will depend on sustained investment, technological innovation, and collaborative efforts across international and commercial stakeholders. If successful, the synthesis of materials in space will not only enable humanity to explore deeper into the cosmos but also deliver transformative benefits to life on Earth.

The exploration of space-based material synthesis represents one of the most transformative advancements in modern material science and engineering. By leveraging the unique conditions of microgravity environments, including reduced buoyancy-driven convection, the absence of sedimentation, and diffusion-dominated mass transport, scientists have unlocked pathways to creating materials with properties and structures unattainable on Earth. From defect-free protein crystals and high-performance alloys to transparent ceramics and advanced polymers, microgravity has emerged as a frontier laboratory where gravitational constraints no longer dictate material behavior. The synthesis of these materials in space has yielded significant advancements in fields such as pharmaceutical development, aerospace engineering, flexible electronics, and biocompatible medical devices.

One of the most critical findings in space material synthesis is the extent to which gravity affects processes like alloy solidification, crystal growth, polymerization, and ceramic sintering. On Earth, these processes are plagued by gravitational influences that cause phase segregation, uneven heat distribution, and structural defects. In microgravity, these effects are minimized or eliminated, resulting in materials with higher structural integrity, improved mechanical strength, tailored thermal properties, and enhanced optical clarity. Experiments conducted aboard platforms such as the ISS, parabolic flights, drop towers, and autonomous satellites have validated these findings across a broad spectrum of materials, including organics, ceramics, alloys, polymers, and crystals. Furthermore, advanced techniques such as EML, containerless processing, emulsion polymerization, and vapor deposition have demonstrated the ability to precisely control material synthesis parameters in ways impossible under terrestrial conditions.

The implications for material science and engineering are profound. Microgravity has allowed researchers to design and produce materials that push the boundaries of what is technologically feasible. For example, protein crystals grown in microgravity exhibit exceptional structural resolution, enabling pharmaceutical companies to design more effective and targeted therapies. Similarly, metallic alloys and amorphous metals synthesized in space display superior mechanical properties, resistance to thermal fatigue, and reduced structural defects, making them ideal for aerospace applications. Transparent ceramics and FGMs produced in microgravity have opened up new possibilities for advanced optics, thermal barriers, and high-performance electronic components. Moreover, polymers synthesized in microgravity have shown unparalleled structural homogeneity, molecular alignment, and tailored thermal and mechanical properties, creating opportunities for next-generation biomedical implants, conductive polymers, and flexible electronics.

On an industrial scale, the commercialization of space-based material synthesis could revolutionize manufacturing practices. Industries ranging from pharmaceuticals and aerospace to electronics and renewable energy stand to benefit from materials produced in space. However, the scalability of these processes for terrestrial applications remains a significant hurdle. Bridging this gap will require innovative approaches, such as hybrid manufacturing systems, AI-driven automation, and advanced simulation technologies, to replicate microgravity effects on Earth or efficiently integrate space-produced materials into global supply chains. Collaborative partnerships between government space agencies, research institutions, and private-sector stakeholders will be essential for addressing these challenges.

The commercialization of space-based material synthesis holds immense economic promise. As the cost of space access continues to decrease, and reusable rocket technologies become standard, the production and commercialization of materials synthesized in space will become increasingly feasible. Industries will soon have access to space-grown pharmaceuticals, high-purity crystals, HEAs, and defect-free semiconductors, unlocking performance gains and efficiency improvements previously thought impossible. Space-based synthesis will also support sustainable space exploration, with ISRU and additive manufacturing reducing reliance on terrestrial supply chains for long-duration missions.

Space-based material synthesis is not merely an academic or experimental pursuit, it represents a paradigm shift in our understanding of material behavior and our approach to material engineering. The unique conditions of microgravity have allowed scientists to transcend the limitations imposed by Earth's gravitational forces, resulting in materials with exceptional properties and capabilities. While challenges related to cost, logistics, scalability, and safety remain, they are not insurmountable. The integration of AI, robotics, additive manufacturing, and real-time monitoring systems will play a pivotal role in addressing these barriers and enabling the large-scale commercialization of space-derived materials.

As humanity stands on the brink of a new era of space exploration and industrialization, the synthesis of materials in space will be at the heart of these efforts. Whether it's creating defect-free protein crystals for life-saving drugs, superalloys for next-generation jet engines, or lightweight polymers for space infrastructure, microgravity research will continue to unlock possibilities that shape both our technological landscape and our place in the cosmos. The pursuit of space-based material synthesis is not just about creating better materials, it's about redefining the limits of innovation, sustainability, and human achievement. As our journey into space continues, these materials will not only enable us to survive and thrive beyond Earth but also revolutionize life here on our home planet.

CHAPTER 4

Microreactors

Revolutionizing Chemical Synthesis Through Precision and Efficiency for Space

Microreactors, also known as microfluidic or lab-on-a-chip systems, are revolutionizing chemical synthesis by offering precise control over reaction parameters, enhanced safety, and improved efficiency. These miniaturized devices enable rapid heat and mass transfer, making them ideal for handling hazardous or highly exothermic reactions. Their modular design allows for easy scalability through parallelization, supporting both small-scale research and industrial-scale production. By integrating synthesis, separation, and analysis into compact systems, microreactors reduce waste, minimize contamination risk, and support process intensification. While challenges such as clogging and fabrication costs persist, advancements in materials science and microfabrication are steadily overcoming these limitations. With applications spanning pharmaceuticals, fine chemicals, and materials science, microreactors represent a transformative step toward more sustainable, agile, and precise chemical processing.

Microreactors, also known as microfluidic devices or lab-on-a-chip systems, have emerged as transformative tools in chemical synthesis, offering unparalleled precision, efficiency, and safety. These devices consist of

DOI: 10.1201/9781003656081-4

miniaturized channels and chambers, typically ranging from micrometers to millimeters in size, where chemical reactions, mixing, and analysis occur. Their design allows precise control over reaction parameters such as temperature, pressure, and flow rates, enabling reactions that are otherwise challenging or hazardous in conventional batch reactors. Microreactors are particularly advantageous for reactions requiring strict control over exothermic behavior, rapid heat transfer, and highly efficient mixing. As industries such as pharmaceuticals, fine chemicals, and materials science increasingly seek faster and safer synthetic processes, microreactors have emerged as indispensable tools for modern chemical engineering and laboratory research.

One of the most significant advantages of microreactors is their exceptional safety profile. Due to the small volumes of reactants involved, even highly exothermic or hazardous reactions can be conducted with minimized risks. This is especially beneficial in handling toxic, flammable, or unstable intermediates, as any accidental release or runaway reaction is contained within the confined space of the microreactor. Additionally, the improved heat and mass transfer properties within microreactors allow reactions to occur under conditions that would be difficult to control in traditional batch reactors. For example, temperature gradients are significantly reduced, and uniform mixing is achieved almost instantaneously, resulting in higher product yields, improved selectivity, and fewer undesired by-products.

Another key advantage of microreactors is their scalability and modularity. Microfluidic devices can be easily scaled up by numbering-up, operating multiple microreactor units in parallel, rather than increasing the size of a single reactor. This approach maintains the same reaction conditions and efficiencies while enabling higher production volumes. Such modular designs also allow researchers to quickly adapt and optimize reaction parameters, facilitating rapid prototyping and iterative experimentation. This flexibility is invaluable in pharmaceutical research, where drug candidates must be synthesized and screened efficiently.

Microreactors also play a crucial role in process intensification, where multiple chemical processes such as synthesis, separation, and analysis can be integrated into a single device. This minimizes the need for intermediate storage and transfer steps, reducing waste, costs, and potential contamination risks. Furthermore, their compact design reduces the physical footprint of chemical synthesis equipment, offering an environmentally friendly and energy-efficient alternative to traditional reactors.

In addition to industrial applications, microfluidic devices have become indispensable in academic research and laboratory studies. Researchers can explore novel reaction pathways, optimize reaction conditions, and study reaction kinetics in real time with unparalleled precision. The integration of advanced analytical tools, such as in-line spectroscopy and chromatography, within microreactors allows continuous monitoring of reaction progress and product formation. This capability enhances both the speed and accuracy of chemical process development.

Despite their numerous advantages, microreactors are not without challenges. Issues such as clogging, limited compatibility with highly viscous fluids, and fabrication costs remain barriers to their widespread adoption. However, ongoing advancements in materials science, additive manufacturing, and microfabrication technologies are addressing these limitations, making microreactors more robust, cost-effective, and adaptable.

Microreactors represent a paradigm shift in chemical synthesis, offering unprecedented control, efficiency, and safety. Their applications span from small-scale laboratory experiments to large-scale industrial production, addressing the growing demand for sustainable, efficient, and precise chemical processes. As research and technology continue to advance, microreactors are poised to play an increasingly central role in the future of chemical engineering, pharmaceutical development, and beyond.

THE ROLE OF MICROREACTORS IN SPACE RESEARCH

- Overview of microreactors as compact, efficient, and controllable reaction systems.
- Importance of synthesizing novel materials in space environments (microgravity or zero gravity).

Microreactors, also known as microfluidic devices or lab-on-a-chip systems, represent a groundbreaking advancement in chemical engineering and material synthesis, offering unparalleled precision, efficiency, and control in conducting chemical reactions. These highly compact systems consist of micro-scale channels, chambers, and reactors, typically etched or molded into materials such as glass, silicon, or polymers. Within these confined spaces, reactants are mixed, heated, and processed with exceptional accuracy, allowing for precise reaction control at a microscopic level. Unlike traditional batch reactors, which rely on large volumes and often face limitations in heat and mass transfer, microreactors enable faster mixing, superior heat exchange, and reduced diffusion limitations, making them ideal for reactions requiring tight control over process variables. Their modular nature allows researchers to scale up synthesis processes by operating multiple units in parallel, a strategy known as numbering-up, rather than increasing the size of a single reactor. This modular design is especially valuable in space environments where payload capacity, weight restrictions, and energy efficiency are critical considerations.

In addition to their physical compactness, microreactors excel in safety and resource efficiency. The small internal volumes significantly reduce the risks associated with hazardous or exothermic reactions, as any accidental runaway or contamination remains confined within a tiny, manageable space. Furthermore, microreactors are capable of operating with minimal amounts of reagents and solvents, which align well with the resource constraints faced in space missions. Their design also allows for in-situ real-time monitoring using integrated sensors and analytical tools, providing immediate feedback on reaction conditions and outcomes. This feedback loop enables precise adjustments during operation, optimizing product quality and minimizing waste. These advantages make microreactors not only valuable tools for laboratory research on Earth but also highly suitable for space-based applications, where every resource, from chemicals to energy, must be carefully conserved.

The importance of synthesizing novel materials in space environments, particularly under microgravity or zero-gravity conditions, cannot be overstated. In terrestrial laboratories, gravity plays a critical role in influencing fluid dynamics, sedimentation, phase separation, and crystallization processes during material synthesis. However, in microgravity, these gravitational effects are largely eliminated, allowing researchers to study and exploit new reaction pathways and material properties that are difficult, if not impossible, to achieve on Earth. For example, in the absence of buoyancy-driven convection currents, crystals can grow with fewer defects, resulting in purer and more uniform crystalline structures. Similarly, sedimentation, which often causes inconsistencies in composite materials during synthesis, is minimized in microgravity, leading to more homogeneous material formation.

Space-based synthesis of organic, inorganic, ceramic, crystalline, metallic, and polymeric materials holds immense promise for various applications, including advanced structural components, electronic devices, biomedical implants, and optical sensors. For instance, high-performance alloys with perfectly distributed microstructures can be produced without the gravitational segregation seen on Earth. Similarly, polymers with highly consistent chain structures and enhanced mechanical or thermal properties can be synthesized under microgravity. Such materials are not only essential for building space habitats, spacecraft shielding, and equipment but also have far-reaching implications for Earth-based industries, driving innovation in aerospace, electronics, healthcare, and energy sectors.

Moreover, the ability to synthesize materials directly in space eliminates the logistical and financial burden of transporting pre-manufactured components from Earth. Instead, raw materials or precursors can be carried in compact forms and synthesized on-demand using microreactor systems aboard spacecraft or space stations. This capability aligns with the vision of sustainable space exploration, where astronauts and scientists can produce critical materials and chemical compounds in situ without relying on continuous resupply missions from Earth.

Microreactors offer an unprecedented opportunity to redefine material synthesis and chemical processing in space. Their compact size, precise control capabilities, and resource efficiency make them ideal for the unique constraints of microgravity environments. As humanity pushes further into space exploration, from long-duration missions to Mars to the establishment of lunar bases and beyond, microreactors will play a pivotal role in enabling self-sustaining, innovative, and efficient material production processes. They are not merely tools for experimentation but essential enablers of a future where space is not only explored but also inhabited and industrialized.

UNIQUE CHALLENGES OF MATERIAL SYNTHESIS IN SPACE

- Absence of convection and sedimentation.
- Controlled mixing and heat transfer in zero-gravity conditions.
- Material purity and precision requirements.

Material synthesis in space presents a unique set of challenges due to the absence of gravitational forces that dominate chemical and physical processes on Earth. In microgravity or zero-gravity environments, fundamental phenomena such as convection, sedimentation, and buoyancy-driven mixing are drastically altered or entirely absent. These changes have profound implications on fluid dynamics, heat transfer, and the quality of synthesized materials. Addressing these challenges is essential for achieving reliable, reproducible, and high-quality material synthesis in space, whether for scientific experiments, structural components, or advanced functional materials.

Absence of Convection and Sedimentation

On Earth, natural convection occurs as warmer, less dense fluids rise while cooler, denser fluids sink, creating continuous flow patterns. These convection currents play a critical role in mixing reactants, dissipating heat, and maintaining uniform concentration gradients within reaction systems. However, in the microgravity conditions of space, convection is largely absent because the buoyancy forces driving fluid motion are negligible. As a result, fluid mixing relies entirely on diffusion and forced flow mechanisms rather than gravitational effects. This absence of convection can lead to uneven distribution of reactants, localized hot spots, and incomplete reactions unless properly managed.

Similarly, sedimentation, which occurs when particles settle under the influence of gravity, is effectively eliminated in space. While this might seem advantageous for preventing phase separation in multiphase systems, it introduces challenges when intentional sedimentation is required for specific processes, such as layered material synthesis or particle deposition. In systems where particle settling is part of the desired outcome, alternative strategies such as precision-controlled flow dynamics or magnetic and electrostatic forces must be employed. The elimination of gravitational sedimentation also complicates the formation of certain composite materials, where controlled phase separation might be essential for achieving the desired structural or functional properties.

To overcome these challenges, microreactors offer a unique advantage. Their small-scale design allows for precise control of reactant flow through microchannels, ensuring adequate mixing through laminar flow patterns and diffusive transport. Furthermore, microfluidic devices can incorporate external forces, such as magnetic or acoustic fields, to induce movement and alignment of particles or fluid layers in the absence of gravity. These engineered solutions help address the challenges posed by the lack of natural convection and sedimentation in space environments.

Controlled Mixing and Heat Transfer in Zero-Gravity Conditions

Efficient mixing and heat transfer are foundational requirements for chemical synthesis and material processing. On Earth, these processes are enhanced by gravitational convection, turbulence, and density-driven gradients. In space, the lack of gravitational forces means that mixing relies solely on diffusion and forced fluid flow mechanisms. Diffusion, however, is an inherently slow process, especially for viscous fluids or reactions requiring rapid and homogeneous mixing. This limitation can result in reaction inefficiencies, poor yield, and non-uniform material properties.

Heat transfer in space is similarly affected by the absence of convection. On Earth, thermal energy is efficiently dissipated through convective heat flow in liquids and gases. In microgravity, heat transfer occurs predominantly through conduction and radiation, both of which are significantly less effective for removing localized heat build-up in reactive systems. This can lead to the formation of thermal hot spots, localized overheating, or incomplete reactions in thermally sensitive materials.

Microreactors address these heat and mixing challenges through their intrinsically high surface-area-to-volume ratios and micro-scale fluid pathways. These devices allow for highly efficient heat exchange across thin reactor walls, preventing thermal gradients and maintaining uniform reaction temperatures. Additionally, microfluidic designs enable precise flow control, where reactants can be mixed rapidly in controlled geometries such as micromixers and split-and-recombine channels. These controlled environments ensure that chemical reactions occur under optimized thermal and mixing conditions, even in the absence of gravitational convection.

Furthermore, microreactors can integrate external actuators such as piezoelectric elements, electromagnetic fields, or ultrasound waves to enhance mixing and heat dissipation actively. These external interventions replicate some of the effects typically achieved through gravitational convection, ensuring that material synthesis proceeds efficiently and predictably in space conditions.

Material Purity and Precision Requirements

In space-based material synthesis, achieving high purity and precision is not just a goal but a necessity. Space environments demand materials with exceptional reliability, minimal defects, and highly controlled properties to withstand the harsh conditions of space missions, such as extreme temperatures, radiation exposure, and mechanical stresses. Whether synthesizing structural materials for spacecraft, electronic components for communication systems, or biomedical materials for onboard healthcare, precision at the molecular and microscopic levels is non-negotiable.

On Earth, gravity-driven processes such as sedimentation or convection can sometimes aid in defect resolution during crystal growth or material formation. However, in microgravity, these natural corrective mechanisms are absent, requiring engineered solutions to maintain consistency and precision throughout the synthesis process. Without precise control over reaction kinetics, phase separations, and growth rates, materials synthesized in space could exhibit structural defects, inconsistent composition, or non-uniform performance characteristics.

Microreactors are uniquely suited to address these requirements. Their small reaction volumes and precisely controlled flow rates ensure that every step of the reaction process can be meticulously monitored and adjusted in real time. Integrated sensors, including spectroscopic tools, thermal monitors, and pressure sensors, allow continuous tracking of reaction parameters, enabling rapid intervention if deviations occur. Additionally, microreactors support the synthesis of high-purity materials through multi-step continuous-flow processing, where intermediate products can be purified and separated inline without requiring manual intervention.

Furthermore, the modular and automated nature of microreactors enables repeatable and reproducible synthesis protocols, reducing variability across production batches. For applications requiring extreme material precision, such as semiconductors, photonic crystals, or specialized polymers, microreactors provide the level of control needed to achieve consistent outcomes across multiple runs, even in zero-gravity conditions.

While the challenges of convection-free mixing, limited heat transfer, and strict material purity requirements are significant in microgravity environments, microreactors offer a powerful solution. Their intrinsically controllable micro-environments, integration with real-time monitoring tools, and adaptability to external force fields provide scientists and engineers with the means to synthesize advanced materials reliably in space. Overcoming these challenges is not merely about solving technical problems but about unlocking new frontiers in space exploration, manufacturing, and self-sustaining human presence beyond Earth.

OBJECTIVES OF SPACE-BASED MICROREACTOR RESEARCH

- Develop advanced organic, inorganic, ceramic, crystalline, metal alloys, and polymer materials.
- Enable sustainable manufacturing capabilities in space environments.

The development and deployment of microreactors in space-based research and manufacturing represent a transformative approach to overcoming the unique challenges of material synthesis in microgravity and zero-gravity environments. Space missions, whether aboard the ISS, lunar bases, or deep-space expeditions, demand innovative strategies to reduce reliance on Earth-bound supply chains, optimize resource utilization, and develop materials with superior properties tailored to the extreme conditions of space. The objectives of microreactor research in space revolve around two primary goals: (1) the development of advanced organic, inorganic, ceramic, crystalline, metal alloys, and polymer materials and (2) enabling sustainable manufacturing capabilities in space environments.

Develop Advanced Organic, Inorganic, Ceramic, Crystalline, Metal Alloys, and Polymer Materials

One of the foremost objectives of space-based microreactor research is to synthesize advanced materials with properties that are either enhanced or unattainable under Earth's gravitational conditions. Different classes of materials, organic compounds, inorganic materials, ceramics, crystalline structures, metal alloys, and polymers, each have unique requirements and applications in space and terrestrial industries. Microgravity offers a distinct environment where fundamental limitations caused by convection, sedimentation, and gravitational phase separation are eliminated, enabling scientists and engineers to explore novel reaction pathways, optimize crystal growth processes, and achieve unprecedented material homogeneity.

- Organic Materials: Organic compounds, including pharmaceutical precursors, bioactive molecules, and lightweight composites, can be synthesized in microreactors with exceptional precision. In space, the absence of sedimentation and phase stratification allows scientists to achieve more consistent molecular structures and purities, which is crucial for pharmaceuticals intended for space missions. Microreactors also facilitate multi-step organic synthesis within a compact and modular platform, enabling the on-demand production of life-supporting chemicals, medicines, and nutritional supplements during long-duration missions.
- Inorganic Materials: Inorganic materials, such as complex oxides, silicates, and phosphates, are critical for advanced electronic components,

coatings, and sensors used in space technologies. In microgravity, microreactors can tightly control the nucleation and growth phases of these materials, preventing structural defects caused by gravitational forces on Earth. These materials can be tailored for use in high-efficiency solar panels, advanced optics, and catalytic systems essential for life-support and energy management aboard spacecraft.

- Ceramic Materials: Ceramics synthesized in space offer unique properties, such as high thermal resistance, electrical insulation, and mechanical durability. The microreactor platform allows precise temperature control and homogeneous mixing of reactants, enabling the synthesis of defect-free ceramic matrices. Applications for space-synthesized ceramics include heat shields, optical ceramics for telescopes, and structural reinforcements for spacecraft components.
- Crystalline Materials: Crystals grown in microgravity are often of significantly higher quality than those grown on Earth due to the absence of convection-driven imperfections. Microreactors provide a controlled environment for crystal nucleation and growth, facilitating the synthesis of advanced semiconductor crystals, photonic materials, and optical-grade structures. These materials are crucial for communication systems, sensor technologies, and optical instruments used in space exploration and remote sensing.
- Metal Alloys: The synthesis of advanced metallic alloys in space is a high-priority objective due to the possibility of achieving homogeneous compositions and fine microstructures free from gravitational segregation. Microreactors enable precise control over reaction temperatures, cooling rates, and mixing conditions, allowing scientists to produce HEAs, shape-memory alloys (SMAs), and corrosion-resistant metals. These materials are indispensable for building spacecraft structures, space station components, and durable tools for extraterrestrial habitats.
- Polymers: Space-based polymer synthesis focuses on high-strength elastomers, conductive polymers, and biocompatible plastics. Microreactors allow the fine-tuning of polymerization reactions, ensuring consistent chain lengths and defect-free structures. Applications of space-synthesized polymers include lightweight structural composites, advanced insulation materials, and biomedical devices for astronauts.

Collectively, the ability to synthesize these advanced materials in space opens doors to innovations in aerospace engineering, medical research, electronics, and energy systems. By leveraging microreactors, researchers can push the boundaries of material science and create substances with superior characteristics that address both space-specific and terrestrial challenges.

ENABLE SUSTAINABLE MANUFACTURING CAPABILITIES IN SPACE ENVIRONMENTS

The second critical objective of space-based microreactor research is to enable sustainable manufacturing capabilities that reduce reliance on Earth-bound supply chains and facilitate self-sustaining operations in space habitats and spacecraft. Space missions are inherently resource-constrained, with every kilogram of payload incurring significant costs. Therefore, on-demand production, recycling, and resource efficiency are paramount for sustainable operations in extraterrestrial environments.

- Resource Efficiency: Microreactors operate with minimal reagent volumes and energy requirements, making them ideal for space-based manufacturing where every resource must be carefully managed. The small volumes involved in microreactors minimize chemical waste, while precise reaction control reduces unnecessary by-products. This level of efficiency aligns perfectly with the sustainability goals of long-duration space missions.
- On-Demand Manufacturing: Space missions often encounter unpredictable needs, from repair materials for equipment failures to specific pharmaceuticals for medical emergencies. Microreactors can be configured for on-demand production of chemicals, specialty materials, and essential compounds, reducing the need for pre-synthesized materials to be transported from Earth. For example, astronauts could synthesize adhesives, sealants, or even specialized lubricants for machinery repair directly onboard.
- In-Situ Resource Utilization (ISRU): Future missions to the Moon, Mars, and beyond rely on harnessing local resources (ISRU) for survival and manufacturing. Microreactors can be integrated with ISRU systems to process raw materials, such as lunar regolith or Martian atmospheric CO_2, into valuable chemicals, fuels, or construction materials. This integration ensures that space habitats can function independently from Earth for prolonged periods.

- Closed-Loop Systems: Sustainable manufacturing in space depends on closed-loop systems, where waste products are recycled and converted back into usable resources. Microreactors can serve as key components in these systems by enabling chemical recycling, purification, and conversion processes with minimal energy inputs. This closed-loop approach ensures minimal waste and maximum resource utilization.
- Scalability and Modularity: Microreactors are inherently modular, meaning multiple units can be operated in parallel to scale production without increasing the complexity of individual reactor systems. This scalability allows astronauts and space engineers to increase manufacturing output dynamically based on mission requirements.

In essence, sustainable manufacturing capabilities powered by microreactors are vital for reducing the dependency on Earth, lowering mission costs, and ensuring long-term survivability and productivity in space habitats. They not only address immediate operational needs but also lay the foundation for future industrialization in space, including potential commercial ventures such as space-based pharmaceutical production, zero-gravity crystal manufacturing, and extraterrestrial mining operations.

The objectives of space-based microreactor research are deeply intertwined with humanity's ambitions for sustainable space exploration, long-duration missions, and even colonization of extraterrestrial environments. The ability to synthesize advanced materials with unparalleled precision and efficiency and to enable resource-conscious, on-demand manufacturing systems represents a monumental step forward in achieving self-sufficient and scalable space operations. These objectives are not just about solving technical challenges, they represent the foundation for a new era of space-based innovation and human expansion beyond Earth.

ADVANTAGES OF MICROREACTORS IN MICROGRAVITY APPLICATIONS

- Controlled Mixing and Reaction Dynamics: Enhanced mixing via microfluidic flow control.
- Precision and Repeatability: Consistent synthesis results under confined, controlled conditions.
- Resource Efficiency: Minimal reagent consumption.

- Real-Time Monitoring and Adjustments: Integration with in-situ analytical tools for real-time quality control.
- Scalability for On-Demand Synthesis: Modular reactor design for parallel processing.

SYNTHESIS OF ORGANIC MATERIALS IN MICROGRAVITY

The synthesis of specialty organic compounds, bioactive molecules, and pharmaceutical precursors in microgravity-enabled microreactors represents a transformative advancement in chemical manufacturing and materials science. Microreactors, with their precise control over reaction parameters and modular scalability, are uniquely suited for space-based synthesis, where the absence of gravity eliminates phase separation, enhances diffusion, and enables highly uniform chemical environments. These conditions result in purer products, improved reaction kinetics, and greater structural consistency, benefits especially critical for pharmaceutical compounds, bioactive proteins, and high-performance organic materials.

In microgravity, microreactors facilitate the efficient synthesis of pharmaceutical precursors and active ingredients with exceptional molecular fidelity, enabling on-demand drug production in space. They also support the formation of complex bioactive molecules and crystals with enhanced therapeutic properties and stability. Additionally, microgravity offers significant advantages for polymerization and composite formation, allowing the development of defect-free, lightweight organic materials for aerospace structures, radiation shielding, flexible electronics, and self-healing spacecraft components.

Eliminating gravitational effects allows for the continuous flow synthesis of complex organic molecules and advanced materials without issues such as stratification or gas bubble coalescence, which often hinder terrestrial processes. This capability is essential for developing sustainable space missions and for improving Earth-based manufacturing through insights gained from space research. As space exploration advances, microreactor-enabled synthesis in microgravity will play a critical role in creating the materials and pharmaceuticals needed to support long-duration missions, while simultaneously driving innovation in terrestrial chemical engineering and biotechnology.

The synthesis of specialty organic compounds, bioactive molecules, and pharmaceutical precursors in microgravity presents a unique opportunity to address several challenges faced in terrestrial laboratories. Organic materials are fundamental building blocks for pharmaceuticals, advanced polymers, biocompatible coatings, and functional materials used in aerospace and medical applications. Space-based microreactors offer a controlled environment where the synthesis of these materials can be conducted with exceptional precision, reproducibility, and efficiency, overcoming many of the limitations posed by Earth's gravity.

In microgravity, organic chemical reactions, including polymerization, complex organic syntheses, and biocatalytic processes, can be fine-tuned to optimize both reaction kinetics and product yield. Specialty organic compounds, such as fine chemicals, organic dyes, or molecular sensors, often require highly controlled reaction conditions to avoid impurities and inconsistencies in structure. Microgravity eliminates gravitationally induced phase separation, ensuring uniformity in chemical compositions and material distribution. This results in higher purity and fewer structural defects, which is crucial for applications where molecular-level precision is required, such as in organic electronics and photonics.

Similarly, bioactive molecules such as proteins, peptides, and enzymes are highly sensitive to their synthesis environments. Microreactors in microgravity can create optimal conditions for protein folding, crystallization, and stabilization, leading to biologically active molecules with enhanced efficacy and stability. For example, proteins synthesized in microgravity have been shown to exhibit more uniform crystalline structures, improving their performance as pharmaceutical agents or biocatalysts. Space-based synthesis of these bioactive molecules holds significant promise for on-demand drug manufacturing in space, where transporting pre-made pharmaceuticals from Earth is impractical due to weight and shelf-life limitations.

Pharmaceutical precursors, the building blocks for essential medicines, can also be synthesized with higher precision in microgravity. Many pharmaceutical compounds rely on multistep organic syntheses that require intermediate purifications and controlled reaction kinetics. Microreactors are particularly well-suited for these multistep processes, allowing sequential reactions to occur seamlessly within a compact device. This capability is essential for producing active pharmaceutical ingredients (APIs) on demand during space missions, reducing dependence on Earth-based supply chains. In scenarios where specific medications are urgently required, microgravity-enabled microreactors could synthesize essential pharmaceutical compounds from simple precursors carried aboard spacecraft.

The ability to synthesize specialty organic compounds, bioactive molecules, and pharmaceutical precursors in space represents a transformative advancement not only for space missions but also for Earth-bound industries. The insights gained from space-based organic synthesis can lead to improved production techniques, purer products, and novel compounds with enhanced properties that are difficult to achieve under terrestrial conditions.

Reaction Mechanisms in Microgravity: Elimination of Gravitational Effects on Phase Separation

One of the most significant advantages of conducting organic synthesis in microgravity is the elimination of gravitational effects on phase separation. On Earth, gravity drives phase separation in multiphase systems, where denser components settle at the bottom, and lighter components rise to the top. This gravitational phenomenon often leads to inhomogeneous reaction environments, uneven mixing, and inconsistent product quality in chemical synthesis. In microgravity, these gravitational effects are essentially neutralized, allowing reactions to proceed in a more uniform and controlled manner.

In traditional chemical reactors, phase separation is a persistent challenge in immiscible liquid–liquid systems, gas–liquid reactions, and emulsions. For example, when synthesizing organic molecules in biphasic systems, gravitational forces can cause the heavier phase to settle, limiting the interfacial area for reaction and reducing overall efficiency. In contrast, microgravity eliminates sedimentation and stratification, ensuring that reactants remain evenly dispersed throughout the reaction vessel. This uniform dispersion enhances mass transfer rates, increases the efficiency of catalytic reactions, and promotes more complete conversions of reactants into desired products.

Additionally, in microgravity, nucleation and growth mechanisms of organic crystals or polymeric structures can be carefully controlled without gravitational disturbances. On Earth, gravity can cause density-driven convection currents within the reactor, leading to variations in temperature and concentration gradients. These inconsistencies often result in defects in crystal lattices or irregular polymer chain structures. In microgravity-enabled microreactors, these disturbances are minimized, allowing for the growth of highly ordered crystalline structures and consistent polymer architectures. This level of precision is particularly critical in applications requiring optically clear organic crystals or high-performance polymer materials.

Microreactors in microgravity further enable enhanced diffusion-driven mixing mechanisms, where reactants are mixed at the molecular level through engineered microchannel designs. In terrestrial batch reactors, gravitational effects can create stagnant zones where reactants are not efficiently mixed, leading to incomplete reactions and the formation of undesired by-products. In contrast, the absence of gravity in microreactors ensures that diffusion becomes the dominant mixing mechanism, facilitating more uniform reactant distribution and improved product yields.

Another crucial benefit of phase-separation-free environments is the ability to synthesize complex organic molecules via multi-step or tandem reactions in a single continuous flow system. In terrestrial conditions, phase separation between intermediate and final products can create bottlenecks in continuous flow synthesis, requiring costly and time-consuming purification steps. In microgravity-enabled microreactors, intermediates remain suspended and reactive within the flow channels, allowing multi-step synthesis processes to proceed seamlessly. This capability reduces waste, lowers energy consumption, and enhances the overall efficiency of chemical synthesis.

Moreover, gas–liquid reactions, such as hydrogenation or oxidation, benefit immensely from the absence of gravity-driven phase separation. On Earth, gas bubbles tend to rise and coalesce, reducing the gas–liquid interfacial area critical for reaction efficiency. In microgravity, gas bubbles remain suspended uniformly within the liquid phase, increasing the surface area for gas–liquid interaction and improving reaction kinetics. This phenomenon has significant implications for processes such as organic hydrogenation reactions and CO_2 conversion into value-added organic chemicals aboard spacecraft.

The elimination of gravitational effects on phase separation in microgravity-enabled microreactors creates a uniform, controlled, and highly efficient reaction environment for organic synthesis. This unique capability allows scientists and engineers to overcome long-standing limitations in terrestrial chemical processes, unlocking new opportunities for producing specialty organic compounds, pharmaceutical precursors, and bioactive molecules with enhanced properties and greater precision. As humanity ventures deeper into space, the ability to synthesize organic materials on-demand using microgravity-enabled microreactors will become an indispensable technology for both space sustainability and cutting-edge terrestrial applications.

Applications of Organic Material Synthesis in Microgravity

The ability to synthesize organic materials in microgravity using microreactors unlocks a range of transformative applications that extend far beyond the confines of traditional terrestrial chemistry. From space-based pharmaceutical production to the development of lightweight organic composites for spacecraft, the unique conditions of microgravity offer advantages in precision, efficiency, and scalability. These applications not only address critical challenges in long-term space exploration but also have the potential to revolutionize industries back on Earth.

Space-Based Pharmaceutical Production

Pharmaceutical production in space represents one of the most compelling applications of microreactor-enabled organic synthesis in microgravity. On long-duration space missions, such as those to Mars or lunar bases, astronauts may require medications for acute or chronic health conditions, injuries, or disease outbreaks. Transporting pre-synthesized pharmaceuticals from Earth is not always feasible due to limited shelf life, storage constraints, and payload weight limitations. Space-based pharmaceutical synthesis, facilitated by microreactors, offers a dynamic and efficient solution to address these challenges.

Microreactors provide the means to synthesize APIs and bioactive compounds with a high degree of precision. In microgravity, the absence of gravitational phase separation allows for more consistent molecular structures, fewer impurities, and better control over reaction kinetics. This precision is particularly crucial for complex drug molecules, where minor deviations in structure can significantly impact therapeutic efficacy. For example, peptide-based drugs, which are notoriously sensitive to reaction conditions, can be synthesized with greater consistency in microgravity using microreactor technology.

Moreover, microreactors enable on-demand pharmaceutical production, where raw chemical precursors can be stored in compact, stable forms and converted into the required medication only when needed. This reduces reliance on large premanufactured stockpiles, which are susceptible to degradation during extended missions. Drugs like antibiotics, painkillers, and personalized medicines tailored to an astronaut's genetic profile can be synthesized in microgravity reactors, ensuring their availability in real time.

Another advantage of space-based pharmaceutical synthesis is the potential to study and produce drugs with enhanced properties. Certain pharmaceutical crystals, such as protein-based therapies and monoclonal antibodies, have been shown to crystallize more uniformly and with fewer defects in microgravity. These improved crystal structures can enhance the bioavailability, stability, and potency of drugs. Insights gained from these space-based studies can also lead to advancements in terrestrial drug manufacturing processes, offering improved therapies for patients on Earth.

Additionally, microreactors can support biocatalytic synthesis pathways, using enzymes or microbial systems to produce complex organic compounds. Such pathways, when optimized in microgravity, can enable the synthesis of high-value pharmaceutical products with minimal energy and chemical waste. This approach aligns with the sustainability goals of space missions, where minimizing by-products and resource consumption is of utmost importance.

Space-based pharmaceutical production using microreactors addresses critical healthcare challenges in space missions while also contributing to advancements in drug manufacturing technologies on Earth. This capability ensures that astronauts remain healthy, resilient, and prepared for the physical and psychological challenges of extended space travel.

Development of Lightweight Organic Composites for Spacecraft

Another key application of organic material synthesis in microgravity lies in the development of lightweight organic composites designed to meet the rigorous demands of spacecraft engineering. Organic composites, including advanced polymers, carbon-fiber-reinforced materials, and nanocomposites, are integral to modern aerospace structures due to their high strength-to-weight ratios, thermal stability, and resistance to harsh environmental conditions.

In microgravity, the synthesis and processing of these organic composites can achieve unparalleled homogeneity, precision, and defect-free properties. On Earth, gravitational forces can lead to phase separation, uneven polymer curing, and void formation during the synthesis and processing of composite materials. These imperfections compromise the mechanical properties, durability, and reliability of the final material. In contrast, microgravity eliminates sedimentation and density-driven defects, allowing the formation of uniform polymer matrices and precise fiber alignments in composite structures.

Microreactors play a crucial role in facilitating the synthesis of these advanced organic materials. They provide a controlled environment for precise polymerization reactions, ensuring consistency in molecular chain lengths and cross-linking densities. Such control is essential for producing composites with predictable mechanical and thermal properties. Additionally, microreactors allow researchers to experiment with novel precursor chemistries and additive formulations that may be too unstable or reactive for traditional synthesis methods on Earth.

One promising application is the production of radiation-shielding organic composites for spacecraft hulls. Space radiation, including cosmic rays and solar particle events, poses a significant threat to both astronauts and electronic equipment. Certain organic materials, such as hydrogen-rich polymers, have excellent radiation-absorbing properties. Microgravity synthesis enables the production of these polymers with optimized structures, maximizing their shielding effectiveness while minimizing material weight.

Furthermore, microgravity facilitates the development of self-healing organic composites that can autonomously repair small cracks or damage caused by micrometeoroid impacts. Microreactors can synthesize polymeric materials containing encapsulated healing agents or embedded catalyst networks that activate upon mechanical stress or exposure to vacuum conditions. These self-healing materials are invaluable for enhancing the longevity and reliability of spacecraft components, reducing the need for costly and time-consuming repairs.

Another critical application is the synthesis of conductive organic polymers for flexible electronics and lightweight power systems in spacecraft. These materials are essential for creating flexible solar panels, lightweight circuitry, and wearable biosensors for astronauts. Microreactors allow precise control over polymer doping and structural modifications, enabling the production of materials with tunable conductivity, flexibility, and durability.

In addition to spacecraft structures, lightweight organic composites are also vital for space habitats and lunar or Martian colonies. They can be used to construct insulating materials, water purification membranes, and oxygen generation components. Microgravity synthesis ensures that these materials meet the stringent performance and durability requirements necessary for long-term survival in extraterrestrial environments.

The insights gained from synthesizing organic composites in space have significant implications for Earth-based aerospace and automotive industries. Space-optimized composite materials often exhibit superior performance characteristics, and lessons learned from microgravity experiments can be translated into improved manufacturing processes and product designs back on Earth.

The synthesis of organic materials in microgravity using microreactors has far-reaching applications in both space-based pharmaceutical production and the development of lightweight organic composites for spacecraft. These advancements address critical operational and logistical challenges in space missions while pushing the boundaries of materials science, engineering, and healthcare innovation. As humanity prepares for long-duration missions to the Moon, Mars, and beyond, the ability to synthesize and utilize these advanced organic materials will become an essential pillar of sustainable space exploration and colonization efforts.

SYNTHESIS OF INORGANIC MATERIALS IN MICROGRAVITY

Inorganic materials synthesized in microgravity environments offer exceptional structural and functional advantages for sensors, electronic devices, and catalytic systems critical to space missions. Free from gravitationally induced convection and sedimentation, microgravity enables the formation of defect-free crystals, uniform thin films, and high-surface-area catalysts with enhanced electronic, optical, and catalytic properties. These advancements support the development of high-precision semiconductor sensors, piezoelectric materials for structural monitoring, optoelectronic components for laser communication, and transparent conductive films for flexible electronics.

Microgravity conditions also facilitate the synthesis of advanced catalytic materials essential for closed-loop life support and energy systems. Inorganic catalysts synthesized in space exhibit optimized pore structures and active site distribution, enhancing performance in applications such as CO_2 reduction, water electrolysis, fuel cells, and waste recycling. Photocatalysts for air purification and surface sterilization similarly benefit from improved efficiency and light-harvesting properties achieved in space.

These innovations not only improve the resilience, efficiency, and autonomy of spacecraft systems but also have far-reaching implications for Earth-based technologies in energy, environmental management, and healthcare. As space exploration advances, micro-gravity-enabled synthesis of inorganic materials will play a pivotal role in supporting sustainable, long-duration missions and accelerating innovation across both extraterrestrial and terrestrial domains.

Complex Oxides, Phosphates, and Silicates: Pioneering Inorganic Material Synthesis in Space

The synthesis of inorganic materials such as complex oxides, phosphates, and silicates in microgravity represents an exciting frontier in material science. These compounds are critical in a wide range of applications, including electronic components, advanced catalysts, sensors, energy storage devices, and structural materials for both terrestrial and space environments. The microgravity environment provides a unique setting to overcome the physical and chemical constraints imposed by Earth's gravitational forces, enabling researchers to achieve unprecedented precision, homogeneity, and control in synthesizing these materials.

Complex oxides are highly versatile materials known for their electronic, magnetic, and optical properties. They are widely used in solid-state batteries, fuel cells, superconductors, and ferroelectric devices. However, synthesizing complex oxides on Earth often leads to structural imperfections caused by uneven crystallization, gravitational phase separation, and convection-driven inhomogeneities. In microgravity, the absence of these gravitational effects enables the formation of uniform oxide structures with fewer defects and more predictable stoichiometry. Microreactors further enhance this precision by allowing tight control over reaction conditions, including temperature gradients, reactant flow rates, and residence times, resulting in complex oxides with optimized performance characteristics for specialized applications.

Phosphates are another critical class of inorganic materials, playing a key role in energy storage (e.g., lithium iron phosphate batteries), catalysis, and optical applications. The synthesis of phosphates typically involves multi-phase reactions that are highly sensitive to temperature and mixing efficiency. In microgravity, phase separation effects are minimized, and reactants remain evenly distributed, ensuring consistent phosphate formation. Microreactors offer the added benefit of continuous flow synthesis, where phosphate precursors can be precisely introduced and reacted under carefully controlled conditions. This setup ensures uniform particle size distribution, high product purity, and minimized by-products, all of which are essential for high-performance phosphate-based materials.

Silicates, known for their thermal and mechanical stability, are extensively used in ceramics, coatings, and structural reinforcements. However, their synthesis is often plagued by challenges such as inhomogeneous nucleation, uneven crystal growth, and impurity entrapment under gravitational influences. In microgravity, the uniform dispersion of reactants and the absence of density-driven sedimentation allow for precise control over silicate crystallization processes. Microreactors facilitate this by enabling controlled introduction of precursors, optimized temperature profiles, and continuous monitoring of reaction kinetics. As a result, space-synthesized silicates exhibit improved structural integrity, thermal stability, and consistency, making them ideal for high-stress applications in spacecraft and space habitats.

The synthesis of these inorganic materials in microgravity not only enhances their structural and functional properties but also opens pathways to creating entirely new classes of materials with novel functionalities. These advancements have significant implications for both space exploration technologies and Earth-based industries, driving innovation in energy storage, environmental sustainability, and advanced manufacturing.

REACTION CONTROL AND CRYSTALLIZATION: PRECISION GROWTH OF INORGANIC CRYSTALS

The growth of high-quality inorganic crystals is a cornerstone of advanced material science, underpinning technologies such as semiconductors, photonic devices, superconductors, and high-precision optical components. On Earth, gravitational forces introduce several challenges during crystallization, including density-driven convection, sedimentation, and uneven growth rates, which often result in defective crystal lattices, impurities, and mechanical stresses within the crystal structures. In contrast, the microgravity environment eliminates these gravitational effects, providing an ideal setting for precision-controlled crystal growth. When combined with the capabilities of microreactors, researchers gain unparalleled control over crystallization processes, enabling the formation of highly uniform and defect-free crystals.

In terrestrial environments, temperature and concentration gradients caused by natural convection currents are major contributors to crystallization defects. In microgravity, where convection is virtually absent, crystals grow more slowly and uniformly, with greater molecular alignment and minimal structural defects. This slow, controlled growth process allows scientists to achieve crystals with higher optical clarity, fewer dislocations, and improved functional properties. Microreactors further enhance this by offering precise control over temperature profiles, precursor delivery, and residence time, all of which are critical parameters in determining crystal quality.

One area where these advancements are particularly evident is in the growth of semiconductor crystals. High-purity crystals are essential for electronic components such as transistors, sensors, and integrated circuits. On Earth, imperfections in semiconductor crystals can cause inconsistencies in electrical conductivity and device performance. In microgravity, semiconductor crystals such as gallium arsenide (GaAs) or silicon carbide (SiC) can be synthesized with exceptional purity and structural homogeneity, leading to higher-performance electronic and optoelectronic devices.

Additionally, optical crystals such as YAG and sapphire (Al_2O_3)-based crystals benefit immensely from microgravity synthesis. These crystals are used in laser technologies, optical lenses, and radiation detectors. Any defects, inclusions, or stresses within the crystal lattice can degrade their optical performance. In space-based microreactors, optical crystals grow in a stress-free and sedimentation-free environment, ensuring superior optical clarity and mechanical strength.

Microreactors also enable the continuous-flow crystallization of inorganic materials, allowing for real-time monitoring and precise control over nucleation and growth dynamics. Integrated sensors within microreactor systems can track parameters such as temperature gradients, pH levels, and reactant concentrations, providing immediate feedback for adjustments. This capability ensures consistent crystallization outcomes, even during long-duration space missions.

Another significant advantage of space-based crystal growth is the ability to study nucleation mechanisms and crystal formation dynamics without gravitational interferences. Insights gained from these studies can lead to the development of new crystallization protocols and methodologies, applicable to both space and terrestrial manufacturing environments. For example, researchers can explore novel inorganic crystal structures or hybrid organic–inorganic frameworks with unique properties suitable for specialized applications such as energy harvesting, quantum computing, and nanotechnology.

Consider that precision crystal growth in space has direct applications in radiation-tolerant materials. Space environments expose materials to intense radiation, which can degrade their structural and functional properties. Crystals synthesized in microgravity can be engineered at the molecular level to exhibit greater resilience to radiation damage, ensuring their reliability in harsh extraterrestrial environments.

Therefore, the reaction control and crystallization of inorganic materials in microgravity-enabled microreactors represent a technological leap in the synthesis of high-performance crystals for a broad range of applications. From semiconductors and optical devices to radiation-tolerant materials, the benefits of microgravity synthesis are poised to drive innovations across space exploration technologies, advanced electronics, and industrial material manufacturing. These advancements not only address the unique requirements of space missions but also hold immense potential for transforming Earth-based industries and scientific research.

Applications of Inorganic Materials Synthesized in Microgravity

The synthesis of inorganic materials in microgravity, including complex oxides, phosphates, and silicates, opens up groundbreaking applications in sensors, electronic materials, and high-performance catalysts essential for space missions. Microgravity eliminates many of the imperfections caused by gravitational forces, enabling the creation of highly uniform, defect-free, and high-purity inorganic materials with enhanced structural and functional properties. These advancements are critical in addressing the extreme operational conditions of space environments, where reliability, efficiency, and precision are non-negotiable requirements.

Sensors and Electronic Materials for Space Missions

Inorganic materials synthesized in microgravity, such as complex oxides, phosphates, and silicates, play a pivotal role in the development of sensors and electronic devices used in spacecraft, satellites, and space habitats. These materials exhibit unique electronic, optical, and magnetic properties that are essential for applications ranging from navigation and environmental monitoring to communication systems and data processing equipment.

Semiconductor-Based Sensors

Semiconductor crystals, including GaAs, SiC, and indium phosphide (InP), are essential components in radiation detectors, light sensors, and photovoltaic cells. In microgravity, these crystals can be synthesized with fewer defects, higher purity, and superior structural uniformity. The elimination of gravitational convection and sedimentation allows for controlled nucleation and crystal growth, resulting in better charge transport properties and enhanced sensitivity. For example, high-precision radiation detectors synthesized in microgravity can more accurately measure cosmic radiation levels, protecting astronauts and sensitive onboard electronics.

Piezoelectric Materials for Spacecraft Monitoring

Certain inorganic compounds, such as lead zirconate titanate (PZT) and other complex oxides, exhibit piezoelectric properties, the ability to generate an electrical charge under mechanical stress. These materials are used in vibration sensors, pressure monitors, and micro-actuators. Microgravity synthesis ensures the production of uniform crystalline structures, improving their sensitivity, durability, and performance. These sensors play a vital role in monitoring spacecraft structural integrity and detecting minute mechanical stresses caused by micrometeoroid impacts.

Optoelectronic Materials for Communication Systems

Advanced inorganic materials, such as YAG crystals and titanium-doped Al_2O_3, are widely used in laser communication systems and optical signal processing devices. The microgravity environment minimizes impurities and structural defects during crystallization, ensuring superior optical clarity and precision. These materials enable high-speed optical communication systems crucial for transmitting vast amounts of data between spacecraft, satellites, and Earth.

Thin-Film Inorganic Materials for Flexible Electronics

Space missions increasingly rely on flexible electronic systems integrated into lightweight, adaptable surfaces. Thin films of silicate-based materials and transparent conducting oxides (e.g., indium tin oxide, ITO) synthesized in microgravity offer improved conductivity, transparency, and mechanical flexibility. These films can be used in foldable solar panels, wearable sensors, and adaptive optical surfaces on spacecraft.

Environmental Monitoring Sensors

Inorganic materials with catalytic and adsorption properties, such as zeolites and metal-organic frameworks (MOFs), can be synthesized with exceptional uniformity in microgravity. These materials are used in sensors for atmospheric monitoring, carbon dioxide detection, and volatile organic compound (VOC) analysis within spacecraft habitats. Their precise structural integrity ensures reliable, long-term performance in closed-loop life support systems.

In essence, inorganic materials synthesized in microgravity enable the production of high-precision sensors and advanced electronic materials that are vital for safety, communication, navigation, and environmental control in space missions. Their superior properties extend mission capabilities, reduce system failures, and ensure the resilience of critical onboard technologies.

High-Performance Catalysts for Onboard Chemical Processes

Catalysts are fundamental to chemical reactions necessary for sustaining life, generating power, and recycling resources in closed-loop systems aboard spacecraft. Inorganic catalysts, such as metal oxides, zeolites, and phosphates, are essential in facilitating chemical processes efficiently, reliably, and with minimal waste. The microgravity environment offers a unique platform for engineering highly effective, defect-free catalytic materials with optimized surface properties and active sites.

Carbon Dioxide (CO_2) Reduction and Oxygen Regeneration

Inorganic catalysts play a crucial role in CO_2 reduction systems, where carbon dioxide exhaled by astronauts is converted into oxygen and other useful compounds. For example, nickel oxide (NiO) and cobalt-based catalysts are commonly used in CO_2 electrolysis systems. In microgravity, these catalysts can be synthesized with high surface area, uniform pore distribution, and optimized active site density, enhancing their efficiency in facilitating CO_2 reduction reactions. This capability is vital for closed-loop life support systems, reducing reliance on oxygen resupply missions from Earth.

Water Electrolysis for Oxygen Generation

Inorganic electrocatalysts, such as iridium oxide (IrO_2) and ruthenium oxide (RuO_2), are used in water electrolysis systems to generate oxygen for breathing and hydrogen for fuel. Microgravity synthesis ensures that these catalysts exhibit uniform crystal structures and high catalytic activity, improving reaction kinetics and energy efficiency. This efficiency is particularly critical in long-duration missions, where resource conservation is a priority.

Fuel Cells for Power Generation

Fuel cells are essential onboard spacecraft for generating electricity from hydrogen and oxygen. Catalysts such as platinum-based nanoparticles and perovskite oxides are widely used in fuel cell electrodes. Microgravity synthesis allows for precise control over particle size, dispersion, and surface morphology, enhancing the catalytic performance and durability of these materials. This advancement improves the efficiency and lifespan of onboard fuel cells, reducing maintenance requirements and resource consumption.

Waste Recycling and Resource Recovery

Inorganic catalysts are also vital in waste processing and resource recovery systems. For example, zeolite-based catalysts can be used to convert waste gases into reusable compounds, while metal oxides facilitate the breakdown of hazardous organic contaminants. In microgravity, these catalysts can be engineered with optimized pore structures and active site accessibility, improving their efficiency in processing waste streams aboard spacecraft.

Propellant and Combustion Catalysis

Spacecraft propulsion systems often rely on catalytic processes for fuel ignition and combustion control. Catalysts such as cerium oxide (CeO_2) and iridium-based compounds are critical in these processes. Microgravity synthesis enables the creation of highly active and thermally stable catalysts capable of withstanding extreme temperature and pressure fluctuations in propulsion systems.

Advanced Photocatalysts for Space Applications

Photocatalysts, such as titanium dioxide (TiO_2) and zinc oxide (ZnO), are used for photochemical water splitting, air purification, and surface sterilization aboard spacecraft. In microgravity, these photocatalysts can be engineered with enhanced light absorption properties and optimized surface areas, improving their efficiency in utilizing solar energy for critical onboard processes.

These materials address the critical challenges of resource recycling, environmental control, energy generation, and chemical processing aboard spacecraft. Beyond their immediate use in space, the technologies and methodologies developed for synthesizing these materials have profound implications for Earth-based industries, including energy, healthcare, and environmental sustainability. As humanity continues to push the boundaries of space exploration, inorganic materials synthesized in microgravity will remain at the forefront of innovation, enabling safer, more efficient, and more sustainable missions to the Moon, Mars, and beyond.

SYNTHESIS OF CERAMIC MATERIALS IN MICROGRAVITY

The synthesis of ceramic materials in microgravity represents a transformative advancement in materials science, enabling the production of high-purity ceramic matrices, superconductors, and advanced composites with exceptional structural and functional properties. On Earth, gravitational forces introduce defects through sedimentation, convection currents, and thermal gradients, compromising material performance. In microgravity, these disruptive effects are minimized, allowing for uniform particle distribution, precise phase transitions, and enhanced control over crystal growth and densification.

Microgravity-enabled microreactors provide the precise thermal and chemical control needed to overcome challenges in ceramic processing, including uniform heat distribution and phase transition stability. These capabilities yield defect-free materials with superior mechanical strength, thermal resistance, and electrical insulation. Applications span from heat-resistant composites for spacecraft shielding to radiation-resistant optical ceramics used in high-power laser systems and advanced sensor technologies.

Ceramics such as SiC, YAG, spinel, and doped superconductors synthesized in microgravity exhibit enhanced thermal stability, optical clarity, and radiation shielding performance. Additionally, advanced ceramic composites benefit from uniform reinforcement bonding and optimized microstructures critical for high-stress environments. The integration of microreactor technology allows real-time process monitoring, precise thermal cycling, and tailored material architectures.

These advancements have profound implications not only for the sustainability and reliability of long-duration space missions but also for Earth-based industries, including aerospace, energy, defense, and healthcare. Microgravity ceramic synthesis marks a pivotal step toward producing next-generation materials with unmatched durability, precision, and functional performance in the harshest environments, both in space and on Earth.

High-Purity Ceramic Matrices, Superconductors, and Advanced Composites

Ceramic materials are highly valued in aerospace, electronics, energy, and structural applications due to their exceptional mechanical strength, thermal stability, electrical insulation, and resistance to wear and corrosion. In space missions, ceramics are essential for thermal protection systems (TPSs), lightweight structural components, radiation shielding, and advanced electronic systems. However, synthesizing high-quality ceramic materials on Earth is often plagued by gravitational effects such as sedimentation, convection currents, and thermal gradients, which introduce structural defects, compositional inconsistencies, and mechanical weaknesses. Microgravity provides an environment where these gravitational effects are minimized or eliminated, allowing for the precision synthesis of high-purity ceramic matrices, superconductors, and advanced composites with superior properties.

High-Purity Ceramic Matrices

In terrestrial ceramic processing, gravitational sedimentation often results in non-uniform particle distribution during suspension-based synthesis. This leads to defects such as density gradients, voids, and poor particle packing, which compromise mechanical strength and thermal resistance. In microgravity, sedimentation is virtually eliminated, ensuring uniform particle distribution and packing density during ceramic precursor preparation. Microreactors enable precise control over reactant delivery, temperature profiles, and reaction kinetics, resulting in ceramic matrices with near-perfect homogeneity and minimal structural defects. Such materials are invaluable in TBCs, aerospace engine components, and insulating tiles for spacecraft.

Superconductors

Ceramic-based superconductors, such as yttrium barium copper oxide (YBCO) and bismuth strontium calcium copper oxide (BSCCO), are critical for advanced technologies, including magnetic propulsion systems, energy storage devices, and precision sensors. The performance of these materials is highly dependent on their crystal structure, phase purity, and defect density. Terrestrial synthesis of superconducting ceramics is limited by gravitational convection and uneven heat distribution during the critical cooling phase, leading to imperfect phase transitions and lattice distortions. In microgravity, these challenges are significantly reduced, enabling uniform nucleation and controlled crystal growth. Microreactors further refine this process by offering precise control over precursor mixing, thermal gradients, and flow rates, resulting in superconductors with enhanced critical temperatures (Tc) and current-carrying capacities.

Advanced Ceramic Composites

CMCs, which combine ceramic materials with reinforcing fibers or particles, are essential for applications requiring high strength, resistance to thermal shock, and durability under extreme conditions. For example, SiC composites are widely used in spacecraft heat shields, structural supports, and high-temperature fuel cells. However, achieving uniform fiber distribution and strong matrix-reinforcement bonding is challenging under Earth's gravitational influence. Microgravity eliminates density-driven phase separation, ensuring uniform reinforcement distribution and optimal bonding at the matrix–reinforcement interface. Microreactors enable the controlled deposition of ceramic precursors onto reinforcing fibers, resulting in highly engineered composite architectures tailored for specific space applications.

The ability to synthesize high-purity ceramic matrices, superconductors, and advanced ceramic composites in microgravity opens new avenues for space exploration technologies, advanced electronics, and sustainable space habitats. These materials not only address the extreme environmental conditions of space but also push the boundaries of material science and engineering on Earth.

Challenges in Ceramic Formation in Zero Gravity

While microgravity offers significant advantages for ceramic material synthesis, it also introduces unique challenges, primarily centered around uniform heat distribution and phase transition control during the processing stages. These factors are crucial in determining the microstructure, mechanical properties, and functional performance of the final ceramic product.

Uniform Heat Distribution

In terrestrial ceramic processing, gravity plays a key role in facilitating convective heat transfer within melts, solutions, or precursors. Convection ensures even temperature distribution across the material, which is critical for uniform particle sintering, phase transformations, and crystal growth. In microgravity, natural convection is virtually absent, and heat transfer occurs predominantly through conduction and radiation. As a result, thermal gradients can form within the material, leading to localized heating, uneven phase distributions, and defects in the final microstructure.

- Thermal Gradients in Microgravity: Without convection, hot and cold zones can persist within the ceramic precursor or melt. This creates inhomogeneous grain growth and phase separation, which can significantly compromise the mechanical strength, porosity, and thermal stability of the final material.

- Solution: Microreactors address this challenge by employing engineered thermal control systems, including precisely calibrated heating elements, integrated temperature sensors, and advanced thermal management protocols. These systems maintain consistent temperature profiles across the reactor chamber, ensuring even particle sintering, crystallization, and bonding.
- Heat Transfer in Multi-Phase Systems: In ceramic composites, reinforcing fibers or particles often have different thermal conductivities than the surrounding ceramic matrix. In microgravity, uneven heat distribution can cause thermal stress and microcracks at the matrix–reinforcement interface.
- Solution: Microreactor designs incorporating microchannel geometries and active temperature modulation systems help mitigate thermal stress and distribute heat uniformly across composite interfaces.

Phase Transition Control

Ceramic materials undergo several critical phase transitions during their synthesis, including crystallization, sintering, and densification. Each of these transitions is highly sensitive to temperature, pressure, and reaction kinetics. In microgravity, the absence of gravitational effects on convection and sedimentation introduces both advantages and new complexities to these transitions.

- Crystallization Dynamics: During ceramic crystal formation, nucleation and growth processes must be tightly controlled to prevent the formation of defects, impurities, or irregular grain sizes. Without gravity, nucleation sites may form randomly, and growth rates can become uneven.
- Solution: Microreactors offer a solution by enabling controlled delivery of precursors, precise thermal cycling, and real-time monitoring of crystallization kinetics. These systems ensure that nucleation events are evenly distributed and crystal growth occurs uniformly across the material.

- Sintering Challenges: Sintering, the process of compacting ceramic particles into a solid mass using heat, relies on precise control of particle packing density and heat application. In microgravity, poor phase control can result in inconsistent densification, trapped gases, and residual porosity in the ceramic matrix.
- Solution: Microreactors allow for dynamic pressure and temperature adjustments during the sintering process, ensuring optimal densification and minimizing gas entrapment. Advanced monitoring technologies, such as in-situ spectroscopy and thermal imaging, provide real-time feedback for process optimization.
- Solid–Liquid Interfaces: In processes involving solid–liquid interfaces, such as melt-casting or vapor deposition, phase boundaries can become unstable in microgravity due to the lack of buoyancy-driven stabilization. This instability can lead to interface deformations and compositional irregularities.
- Solution: Microreactors employ controlled flow systems and external stabilization techniques, such as electromagnetic or acoustic fields, to maintain stable solid–liquid interfaces.

The synthesis of high-purity ceramic matrices, superconductors, and advanced ceramic composites in microgravity represents a significant technological advancement in aerospace, electronics, and sustainable space exploration. While uniform heat distribution and phase transition control remain significant challenges, the integration of microreactor technology offers innovative solutions through precise thermal management, advanced monitoring systems, and real-time process adjustments. These advancements not only address the extreme demands of space environments but also drive innovation in terrestrial industries, paving the way for next-generation ceramic materials with unparalleled structural integrity and performance capabilities. As humanity ventures deeper into space, ceramic materials synthesized in microgravity will remain critical for mission success, safety, and sustainability.

Applications of Ceramic Materials Synthesized in Microgravity

The synthesis of high-purity ceramic matrices, superconductors, and advanced composites in microgravity unlocks a range of transformative applications in spacecraft shielding and optical sensor systems. Ceramics, known for their exceptional thermal stability, mechanical strength, electrical insulation, and optical clarity, are indispensable in the harsh and unforgiving environment of space. The unique conditions of microgravity, combined with the precision control offered by microreactor technology, enable the production of ceramics with unparalleled purity, uniformity, and defect-free structures, significantly enhancing their functionality and performance.

Heat-Resistant Materials for Spacecraft Shielding

Spacecraft are subjected to extreme thermal environments during atmospheric re-entry, propulsion maneuvers, and exposure to solar radiation and cosmic rays. Heat-resistant ceramic materials are essential for maintaining the structural integrity and safety of spacecraft and their occupants under these conditions. Materials such as SiC, alumina (Al_2O_3), zirconia (ZrO_2), and hafnium carbide (HfC) are widely used for thermal shielding due to their high melting points, low thermal conductivity, and resistance to thermal shock.

Thermal Protection Systems

Heat-resistant ceramic composites form the core of TPSs, which protect spacecraft from the extreme heat generated during re-entry into planetary atmospheres. These materials need to withstand temperatures exceeding 2,000°C (3,632°F) without deforming, melting, or degrading. On Earth, gravitational forces during ceramic synthesis can result in voids, uneven particle distribution, and inconsistent densification, which compromise their thermal resistance. In microgravity, these defects are minimized, allowing for the production of high-density, uniform ceramic matrices with optimized thermal properties.

- Microgravity Advantage: In microgravity, ceramics can be synthesized with uniform particle packing, precise phase transitions, and fewer microstructural defects. These qualities improve their ability to dissipate heat evenly and resist thermal shock during rapid heating and cooling cycles.
- Application Example: Spacecraft such as NASA's Orion and SpaceX's Starship rely on ceramic-based heat shields to endure the intense heat of atmospheric re-entry. Space-synthesized ceramics with optimized porosity, enhanced heat resistance, and reduced thermal expansion mismatches provide longer-lasting and more reliable protection.

Ablative Heat Shields

Certain ceramic composites are designed as ablative heat shields, where the material gradually vaporizes during re-entry, carrying away heat energy. Materials such as reinforced carbon-carbon (RCC) composites benefit immensely from microgravity synthesis, where phase uniformity, fiber alignment, and interfacial bonding can be carefully controlled, ensuring predictable ablation rates and thermal stability.

Radiation Shielding

In addition to thermal management, ceramic materials are also effective for radiation shielding. Ceramics enriched with boron or hafnium compounds have exceptional neutron absorption and radiation attenuation properties. Microgravity synthesis enables the creation of ceramics with optimized atomic arrangements and minimized structural defects, enhancing their radiation shielding capabilities.

So the ability to synthesize heat-resistant ceramic materials in microgravity results in spacecraft shielding systems that are lighter, more durable, and capable of withstanding extreme thermal and radiation stresses, significantly improving spacecraft safety and mission reliability.

Optical Ceramics for Advanced Sensor Systems

Optical ceramics, such as YAG, magnesium aluminate spinel ($MgAl_2O_4$), and Al_2O_3, play a critical role in space-based sensor systems, high-power laser technologies, and optical windows. These materials are prized for their exceptional optical transparency, resistance to radiation damage, and mechanical strength. However, synthesizing high-quality optical ceramics on Earth is challenging due to gravitational effects, which can cause density gradients, inclusion defects, and non-uniform crystallization.

Precision Optical Sensors

Space missions rely heavily on optical sensors for navigation, environmental monitoring, astronomical observations, and remote sensing of planetary surfaces. Optical ceramics are integral components of laser systems, spectrometers, and camera lenses used in these sensors. In terrestrial synthesis, gravitational convection can introduce impurities and uneven crystallization, which affect optical clarity, light transmission efficiency, and precision alignment of crystal lattices.

- Microgravity Advantage: In microgravity, optical ceramics can be grown with minimal inclusions, fewer defects, and highly uniform crystalline structures, ensuring maximum transparency and light transmission. Microreactor technology further enhances this precision by allowing continuous control over crystal growth kinetics and thermal gradients.
- Application Example: YAG crystals synthesized in microgravity can be used in high-energy laser systems for satellite communication or as gain media in laser amplifiers for scientific instruments.

Radiation-Resistant Optical Windows

Optical windows made from ceramics such as Al_2O_3 or spinel are used in spacecraft observation ports, sensor covers, and radiation shielding for optical instruments. These materials need to maintain optical clarity and mechanical integrity under prolonged exposure to cosmic radiation, micrometeoroid impacts, and temperature fluctuations.

- Microgravity Advantage: In microgravity, the absence of sedimentation and convection ensures uniform lattice structures and reduces structural weaknesses caused by inclusions or dislocations. These materials can better resist radiation damage and temperature-induced stress, ensuring long-term operational stability.
- Application Example: Optical ceramic windows are critical for systems like the Hubble Space Telescope, Mars rovers' cameras, and satellite-based Earth observation sensors.

Laser Communication Systems

In the realm of space-based laser communication, optical ceramics are indispensable components of laser transmitters, receivers, and optical amplifiers. Materials such as neodymium-doped YAG (Nd:YAG) or erbium-doped ceramics enable high-energy laser emissions used in secure, long-distance data transmission between spacecraft and ground stations.

- Microgravity Advantage: In space-based microreactors, these doped optical ceramics can be synthesized with precise dopant distributions, minimal defects, and uniform crystal growth. This results in higher optical efficiency and reduced signal loss, enabling more robust and energy-efficient communication systems.

Infrared and Ultraviolet Optical Sensors

Infrared (IR) and ultraviolet (UV) sensors are critical for thermal imaging, atmospheric monitoring, and planetary exploration. Ceramics such as zinc selenide (ZnSe) and magnesium fluoride (MgF_2) are widely used in these sensors. Space synthesis allows for the creation of high-purity optical ceramics with tailored absorption and transmission properties, enhancing the performance of IR and UV systems in space environments.

The synthesis of heat-resistant ceramic materials and optical ceramics in microgravity represents a technological leap in supporting space exploration, satellite operations, and scientific research missions. Heat-resistant ceramics offer improved shielding against extreme thermal and radiation conditions, while optical ceramics enable high-precision sensor technologies and robust communication systems. The defect-free structures, enhanced purity, and superior performance of microgravity-synthesized ceramics address long-standing limitations of terrestrial processing methods. These advancements not only ensure mission reliability, safety, and sustainability in space but also have far-reaching applications in Earth-bound industries, including aerospace, electronics, energy, and healthcare. As humanity continues to expand its presence in space, ceramic materials synthesized in microgravity will remain foundational for cutting-edge technologies and sustainable exploration beyond Earth.

SYNTHESIS OF CRYSTALLINE MATERIALS IN MICROGRAVITY

The development of precision optical devices for space telescopes such as Hubble and James Webb Space Telescope (JWST) and future missions like Large UV/Optical/IR Surveyor *(LUVOIR) depends on the availability of high-purity optical crystals with exceptional clarity, uniformity, and radiation resistance. Microgravity synthesis offers a unique environment to grow defect-free materials such as calcium fluoride, magnesium fluoride, sapphire, and ZnSe, which are critical for lenses, mirrors, spectrometers, and IR/UV sensors. By eliminating gravitational effects like sedimentation and convection, microgravity enables the formation of optical crystals with consistent refractive indices and minimal inclusions, resulting in superior light transmission and imaging performance.*

These advanced materials enhance the durability and resolution of telescopic instrumentation, allowing scientists to observe deep-

space phenomena with unprecedented precision. Applications include radiation-resistant components for long-duration missions, high-transmission IR and UV sensors for celestial imaging, and adaptive optics systems with flawless reflective surfaces. As space exploration advances, microgravity-grown optical materials will play a vital role in enabling more accurate astronomical observations and supporting robust, high-performance space-based communication and research systems.

Crystalline materials, including semiconductor crystals, photonic crystals, and optical-grade materials, play a foundational role in advanced technologies used in electronics, communication systems, photonics, and optical applications. The structural and functional performance of these materials depends heavily on the precision of their crystal lattices, uniformity, and purity. However, crystal synthesis on Earth is frequently hindered by gravitational effects, such as sedimentation, density gradients, convection currents, and buoyancy-driven forces, which disrupt uniform crystal growth and lead to defects, inclusions, and structural irregularities. In microgravity, these gravitational influences are largely eliminated, creating an ideal environment for the controlled growth of defect-free crystals with highly ordered structures. When paired with microreactor technology, scientists can achieve an unprecedented level of control over crystallization kinetics, nucleation processes, and growth dynamics, resulting in materials with superior optical, electronic, and mechanical properties.

Semiconductor Crystals

Semiconductor crystals, such as silicon (Si), GaAs, InP, and cadmium telluride (CdTe), are essential for manufacturing microchips, transistors, photovoltaic solar cells, sensors, and high-frequency communication devices. The efficiency and reliability of these devices are heavily influenced by the purity and structural integrity of the semiconductor crystals used in their fabrication.

On Earth, gravity induces thermal convection currents, sedimentation, and density-driven segregation during crystal growth, leading to defects, dislocations, and impurities in the resulting crystal lattice. These imperfections can reduce the charge carrier mobility, electronic conductivity, and overall device performance of semiconductor materials. In microgravity, these gravitational effects are eliminated, allowing the homogeneous distribution of dopants and reactants, uniform temperature profiles, and defect-free crystal growth.

- Controlled Doping in Space: Doping semiconductor crystals with precise concentrations of elements such as phosphorus, boron, or arsenic is critical for tuning their electrical properties. In microgravity, dopants can be distributed more evenly, minimizing concentration gradients and impurity clustering, which are common challenges in terrestrial crystal growth.
- Photovoltaic Applications: Space-grown semiconductor crystals, such as GaAs-based solar cells, exhibit higher efficiency and longer operational lifespans compared to their Earth-grown counterparts. These advancements are essential for developing high-efficiency solar panels for spacecraft and satellites, where energy reliability is critical.

The integration of microreactors in semiconductor crystal synthesis enhances the precision of these processes by enabling real-time control over precursor flow rates, reaction kinetics, and thermal gradients. This ensures reproducible and defect-free crystal growth, allowing semiconductor materials to meet the stringent quality standards required for space and terrestrial electronic applications.

Photonic Crystals

Photonic crystals are materials with periodic optical nanostructures that influence the propagation of light, enabling applications in optical communication, photonic circuits, lasers, and high-efficiency light-emitting diodes (LEDs). These materials rely on precisely ordered structures at the nanometer scale to manipulate light in specific ways, such as creating photonic band gaps that prevent light from propagating in certain directions.

On Earth, synthesizing photonic crystals with uniform periodic structures is a significant challenge due to gravitational convection, sedimentation, and thermal gradients, which disrupt the formation of consistent patterns at the nanoscale. In microgravity, these effects are significantly reduced, allowing precise self-assembly of photonic crystal structures with minimal defects and variations.

- Enhanced Light Manipulation: Space-grown photonic crystals exhibit superior optical properties, lower scattering losses, and higher structural uniformity, enhancing their performance in laser systems, optical switches, and wavelength-selective mirrors.
- Integration into Photonic Devices: Uniform photonic crystals synthesized in microgravity are better suited for on-chip photonic circuits, enabling faster and more energy-efficient optical data processing systems for space and Earth applications.

Microreactors further optimize the synthesis of photonic crystals by allowing precise control over precursor injection, reaction rates, and thermal conditions. Real-time monitoring systems integrated into microreactors ensure that nucleation and self-assembly processes occur uniformly, minimizing structural irregularities and enhancing the final product's performance.

Optical-Grade Materials

Optical-grade crystals, such as YAG, Al_2O_3, lithium niobate ($LiNbO_3$), and calcium fluoride (CaF_2), are essential for applications in high-power laser systems, optical lenses, fiber-optic communication, and radiation-tolerant windows for spacecraft and satellites. The optical performance of these materials is directly tied to their structural clarity, absence of defects, and homogeneity.

On Earth, gravitational effects during crystal growth introduce imperfections, inclusions, and residual stresses, which degrade optical properties, including light transmission efficiency, refractive index uniformity, and mechanical strength. In microgravity, these defects are minimized, resulting in crystals with higher optical clarity, improved thermal stability, and superior mechanical properties.

- Laser Applications: Space-grown YAG crystals, doped with elements such as neodymium (Nd) or erbium (Er), are used as laser gain media in high-power lasers for communication, ranging, and scientific research. Microgravity-grown crystals exhibit fewer optical losses and higher energy conversion efficiencies.
- Optical Windows and Lenses: Al_2O_3 and CaF_2 crystals synthesized in microgravity are used in space observation windows, telescopic

lenses, and precision optical sensors. These materials maintain clarity and mechanical strength even after prolonged exposure to cosmic radiation and extreme temperatures.

Microreactors provide the level of precision and control required for growing optical-grade materials in space. By enabling dynamic temperature modulation, controlled precursor delivery, and real-time monitoring, microreactors ensure defect-free growth and uniform doping across large crystal volumes.

The Role of Microreactors in Crystallization Kinetics for Uniform Nucleation and Growth Control

The success of crystalline material synthesis in microgravity depends heavily on the kinetic control of nucleation and crystal growth processes. Microreactors are uniquely equipped to address these challenges, providing a highly controlled environment where reaction parameters can be dynamically adjusted in real time.

- Nucleation Control: In crystallization, nucleation is the initial and most critical phase. If nucleation is uncontrolled, it can result in multiple nucleation sites, uneven crystal growth, and structural defects. Microreactors allow scientists to carefully regulate reactant concentrations, temperature gradients, and supersaturation levels to achieve single nucleation events or controlled multiple nucleation as required.
- Growth Kinetics: Crystal growth depends on the availability of reactants, diffusion rates, and temperature stability. In microgravity, natural convection is absent, and reactant delivery relies on diffusion. Microreactors compensate for this by enabling laminar flow mixing and precision precursor injection, ensuring consistent reactant availability throughout the growth process.
- In-Situ Monitoring: Integrated sensors in microreactors, such as spectroscopy probes and thermal cameras, provide real-time data on crystallization kinetics. This enables scientists to adjust parameters on the fly, ensuring optimal growth conditions are maintained throughout the process.
- Scalability: Microreactors offer a modular and scalable platform for crystal growth. Multiple microreactors can be operated in parallel, allowing increased production while maintaining the same level of precision and consistency.

The synthesis of semiconductor crystals, photonic crystals, and optical-grade materials in microgravity, combined with the precision control offered by microreactor technology, represents a paradigm shift in advanced material science and space manufacturing. These materials are critical for electronics, photonics, and optical applications, both in space missions and Earth-based industries. The controlled nucleation, uniform growth, and defect-free structures achieved in microgravity using microreactors enable higher efficiency, longer operational lifespans, and enhanced performance across diverse applications. As humanity ventures further into space, crystalline materials produced in microgravity will play a foundational role in supporting sustainable space exploration, advanced communication systems, and groundbreaking scientific research.

Applications of Crystalline Materials Synthesized in Microgravity

The synthesis of semiconductor crystals, photonic crystals, and optical-grade materials in microgravity has paved the way for groundbreaking advancements in advanced communication systems and precision optical instrumentation. The absence of gravitational convection, sedimentation, and uneven thermal gradients in microgravity enables the creation of high-purity, defect-free crystalline materials with properties far surpassing their Earth-grown counterparts. These superior materials are critical for ensuring reliability, efficiency, and durability in technologies that support space exploration, scientific discovery, and interplanetary communication.

Advanced Communication Devices for Space Applications

Reliable and high-speed communication systems are the backbone of modern space exploration and satellite operations. These systems depend on semiconductor crystals, photonic materials, and optical-grade components for tasks ranging from data transmission and signal amplification to real-time communication between spacecraft and Earth. Microgravity-synthesized crystalline materials provide the structural and optical precision required for efficient signal transmission, reduced energy losses, and enhanced system performance.

Semiconductor Crystals for Communication Electronics

Space missions rely heavily on high-frequency semiconductors, such as GaAs and InP, for radio frequency (RF) amplification, signal processing, and data transmission. These materials serve as the foundation for transceivers, signal amplifiers, and data relays in spacecraft and satellites.

- Microgravity Advantage: On Earth, gravitational convection and sedimentation often introduce defects, impurities, and dopant inconsistencies in semiconductor crystals, leading to signal noise, energy inefficiencies, and reduced lifespan of communication systems. In microgravity, these effects are mitigated, enabling the growth of high-purity, homogeneously doped semiconductor crystals with superior electronic properties.
- Application Example: Microgravity-grown GaAs-based power amplifiers are used in satellite transponders to boost weak signals for long-distance communication, reducing energy losses and improving signal integrity over vast distances.

Photonic Crystals for Optical Signal Processing

Photonic crystals are essential for wavelength filtering, signal routing, and data modulation in optical communication networks. These periodic nanostructures are designed to manipulate light with extreme precision, allowing for high-speed data transmission with minimal interference or loss.

- Microgravity Advantage: On Earth, gravitational effects cause imperfections in the periodic structure of photonic crystals, leading to light scattering, reduced efficiency, and performance inconsistencies. In microgravity, these imperfections are minimized, resulting in uniform periodic structures and precise photonic bandgaps.
- Application Example: Space-grown photonic crystals are used in satellite laser communication systems, enabling secure, high-bandwidth data transfer between spacecraft and ground stations with reduced energy consumption.

Optical-Grade Materials for Laser Communication Systems

Laser communication systems, or free-space optical communication, use highly focused laser beams to transmit data across vast distances. Optical-grade crystals, such as Nd:YAG, serve as the gain media in solid-state lasers, which are integral to these systems.

- Microgravity Advantage: In microgravity, Nd:YAG crystals can be synthesized with fewer inclusions, reduced scattering centers, and optimized dopant distributions, enhancing their optical clarity, energy efficiency, and power output.
- Application Example: Microgravity-grown Nd:YAG lasers are used in inter-satellite laser links (ISLs), providing secure, high-speed communication channels between orbiting satellites, which is essential for data relays, navigation, and Earth observation missions.

Reliability and Longevity in Space Systems

Space communication devices must endure cosmic radiation, extreme thermal fluctuations, and mechanical stresses. Crystals synthesized in microgravity demonstrate enhanced radiation resistance, thermal stability, and structural durability, making them ideal for long-duration space missions and reducing the need for frequent maintenance or replacements.

The crystalline materials grown in microgravity provide the precision, purity, and performance required for advanced communication systems in space. They enable faster data transmission, higher signal integrity, and more efficient energy use, supporting critical space applications such as deep-space exploration, satellite operations, and interplanetary communication networks.

Precision Optical Devices for Telescopic Instrumentation

Modern space telescopes, including Hubble, JWST, and future missions like Large UV/Optical/IR Surveyor (LUVOIR), rely on precision optical devices made from high-purity crystalline materials for image capture, light analysis, and deep-space observations. These devices require flawless optical properties, exceptional clarity, and resistance to radiation and thermal stress, qualities that are enhanced through microgravity synthesis.

High-Purity Optical Crystals for Space Telescopes

Optical-grade materials, such as CaF_2, Al_2O_3, and MgF_2, are used in telescope lenses, mirrors, and spectrometers. These materials must transmit light across specific wavelengths without distortion or scattering to ensure precise astronomical observations.

- Microgravity Advantage: On Earth, gravitational forces can cause density gradients, inclusion defects, and refractive index inconsistencies in optical crystals. In microgravity, these effects are mitigated, resulting in defect-free crystals with uniform refractive indices and superior optical transparency.

- Application Example: Space-grown CaF_2 crystals are used in spectrographs and UV imaging systems aboard telescopes, allowing astronomers to study distant celestial objects with unprecedented clarity.

Infrared and Ultraviolet Sensors for Telescopes

IR and UV sensors rely on optical-grade materials such as ZnSe and $LiNbO_3$ for capturing and analyzing electromagnetic radiation from distant cosmic sources.

- Microgravity Advantage: The precise control of crystallization kinetics in microgravity results in crystals with fewer absorption defects and higher transmission efficiency across specific IR and UV wavelengths.
- Application Example: High-purity ZnSe crystals are used in infrared sensors for telescopes, enabling scientists to study celestial phenomena such as black holes, nebulae, and exoplanets in unprecedented detail.

Radiation-Resistant Optical Components

Space telescopes must endure constant exposure to cosmic radiation and high-energy particles, which can degrade optical components over time. Crystals synthesized in microgravity exhibit enhanced radiation tolerance due to their uniform atomic structures and minimal lattice defects.

- Application Example: Radiation-resistant Al_2O_3 lenses are used in observation ports and optical sensors for long-duration missions, ensuring consistent performance even after prolonged radiation exposure.

Precision Mirrors and Adaptive Optics

Space-based telescopes use advanced optical materials for primary mirrors and adaptive optical systems, which correct distortions caused by atmospheric turbulence and misalignments.

- Microgravity Advantage: Mirrors made from space-grown optical crystals exhibit flawless reflective surfaces and minimal scattering effects, improving image resolution and clarity.

- Application Example: Adaptive optical systems incorporating microgravity-grown optical crystals are used to capture sharp, high-resolution images of distant galaxies and celestial phenomena.

The synthesis of semiconductor crystals, photonic crystals, and optical-grade materials in microgravity represents a critical advancement in supporting advanced communication systems and precision telescopic instrumentation. These applications rely on the superior optical clarity, defect-free structures, and enhanced durability of microgravity-grown crystalline materials. By enabling faster, more reliable communication networks and more precise astronomical observations, these technologies are instrumental in driving humanity's space exploration, scientific discovery, and interplanetary missions. As space missions grow increasingly ambitious, microgravity-synthesized crystalline materials will remain essential for the next generation of space-based infrastructure and scientific research platforms.

SYNTHESIS OF METAL ALLOYS IN MICROGRAVITY

Density-driven separation during alloy solidification is a major challenge in terrestrial metallurgy, leading to non-uniform mechanical properties, localized weaknesses, and performance inconsistencies. In microgravity, the absence of significant gravitational forces eliminates this segregation, allowing alloy constituents to remain uniformly distributed throughout cooling and solidification. This results in homogeneous phase distributions, consistent grain structures, and enhanced mechanical integrity. Microreactor technology further enhances this process by providing precise control over precursor delivery, mixing, and cooling rates, enabling the synthesis of high-performance alloys with exceptional reliability.

The microgravity-enabled production of high-entropy alloys, SMAs, and lightweight metallic composites marks a transformative leap in materials engineering. These advanced materials, synthesized without density-related defects, are ideally suited for demanding applications in aerospace, space exploration, and next-generation terrestrial technologies. As microgravity research and microreactor integration continue to evolve, they promise to unlock new possibilities in creating alloys that are lighter, stronger, and more resilient than ever before.

The synthesis of metal alloys in microgravity offers transformative potential for producing HEAs, SMAs, and lightweight metallic composites with properties superior to their terrestrial counterparts. Alloys are essential materials in aerospace, electronics, construction, and biomedical industries due to their high strength-to-weight ratios, corrosion resistance, and tailored mechanical properties. However, gravitational forces on Earth impose limitations on alloy synthesis, including density-driven separation, convection currents, and inconsistent solidification patterns, which result in inhomogeneous distribution of alloying elements, structural defects, and residual stresses. In contrast, the microgravity environment minimizes these effects, allowing for the creation of uniform, defect-free alloys with optimized microstructures and mechanical properties.

Microgravity eliminates gravity-driven buoyancy and sedimentation effects during alloy solidification, leading to homogeneous mixing of metallic elements and uniform phase distributions. When combined with microreactor technology, which allows precise control over reaction kinetics, precursor delivery, and cooling rates, researchers can synthesize alloys with unprecedented consistency, purity, and tailored properties.

High-Entropy Alloys (HEAs)

High-entropy alloys (HEAs) are a class of metallic materials composed of five or more principal elements mixed in near-equiatomic proportions. Unlike conventional alloys, which rely on a dominant base metal, HEAs derive their exceptional properties from their multi-element composition and highly stable solid-solution structures. These alloys exhibit high strength, excellent corrosion resistance, outstanding thermal stability, and superior fatigue resistance, making them ideal for applications in spacecraft structures, engine components, and thermal shielding systems.

On Earth, gravitational effects during HEA solidification can result in density segregation, phase separation, and inconsistent elemental distributions. These imperfections compromise the alloy's performance, especially in applications requiring uniform mechanical strength and thermal conductivity.

- Microgravity Advantage: In microgravity, the absence of density-driven segregation ensures that all constituent elements are evenly dispersed throughout the alloy matrix during solidification.

This leads to homogeneous microstructures, consistent elemental distributions, and reduced lattice defects.

- Application Example: Space-grown HEAs can be used in spacecraft frames, propulsion systems, and high-temperature shielding, where durability and resistance to extreme conditions are paramount.

Microreactors provide additional control over alloy composition, reaction kinetics, and cooling profiles, ensuring reproducible results and the ability to experiment with novel HEA formulations.

Shape-Memory Alloys

Shape-memory alloys (SMAs), such as nickel-titanium (NiTi) alloys, are known for their ability to recover their original shape after deformation when exposed to specific temperatures. This property, known as the shape-memory effect, along with pseudoelasticity, makes SMAs invaluable in aerospace, robotics, and biomedical engineering. They are used in deployable space structures, self-repairing spacecraft components, and adaptive mechanisms that respond to environmental changes.

On Earth, SMA synthesis is often compromised by thermal gradients, inconsistent cooling rates, and density-driven segregation, which can introduce grain boundaries, inclusions, and impurities in the final material. These imperfections degrade the alloy's shape-memory performance and reduce its fatigue resistance.

- Microgravity Advantage: In microgravity, SMAs can be synthesized with uniform thermal gradients, even elemental distribution, and consistent phase transformations across the entire alloy. This results in materials with predictable and repeatable shape-memory behavior, enhanced fatigue life, and improved mechanical performance.
- Application Example: Space-grown SMAs can be used in deployable antenna systems, self-actuating spacecraft panels, and temperature-regulated adaptive components, where reliability and long-term performance are critical.

Microreactors enable fine-tuned control over phase transformations, elemental diffusion rates, and cooling processes, ensuring that the alloys achieve the desired crystalline phase structures and performance characteristics.

Lightweight Metallic Composites

Lightweight metallic composites, including aluminum-magnesium alloys, titanium composites, and magnesium-lithium alloys, are widely used in aerospace engineering due to their high strength-to-weight ratios, corrosion resistance, and structural efficiency. These materials are essential for spacecraft hulls, structural supports, fuel storage tanks, and lightweight mechanical components.

On Earth, the synthesis of metallic composites is plagued by density segregation, uneven distribution of reinforcing particles, and phase inhomogeneities, which result in weak points and inconsistent mechanical properties across the material.

- Microgravity Advantage: In microgravity, metallic composites can be synthesized with uniform particle dispersion, homogeneous phase distributions, and minimal porosity. The absence of gravitational sedimentation ensures that reinforcing particles (e.g., ceramic particulates or carbon fibers) remain evenly distributed within the metal matrix during solidification.
- Application Example: Space-grown lightweight composites are ideal for aerospace structures, cryogenic storage tanks, and impact-resistant spacecraft shields, where reduced weight and enhanced strength directly improve mission efficiency and payload capacity.

Microreactors further enhance the synthesis of metallic composites by providing dynamic control over precursor delivery, reaction temperatures, and particle dispersion mechanisms, ensuring consistent microstructures and optimal bonding at the matrix-reinforcement interfaces.

Improved Alloy Homogeneity in Microgravity

Alloy homogeneity, the uniform distribution of alloying elements within a metal matrix, is critical for achieving consistent mechanical properties, thermal conductivity, and resistance to fatigue and corrosion. On Earth, achieving alloy homogeneity is challenging due to density-driven segregation, uneven diffusion rates, and gravitational convection currents during the solidification process.

In microgravity:

- Elimination of Density-Driven Segregation: The absence of buoyancy and sedimentation ensures that heavier alloying elements do not sink nor do lighter elements rise during cooling and solidification.

This prevents the formation of composition gradients and localized concentration zones within the alloy.

- Even Elemental Distribution: In microgravity, diffusion becomes the primary driver for elemental distribution, allowing alloying elements to disperse evenly across the matrix without gravitational interference.
- Uniform Microstructure: Controlled solidification in microgravity ensures consistent grain size and orientation, minimizing defects such as voids, inclusions, and grain boundary mismatches.

Microreactors enhance this process by providing precise temperature control, uniform precursor mixing, and real-time monitoring of alloy solidification dynamics. This ensures alloys exhibit consistent mechanical strength, fracture resistance, and thermal stability across the entire material.

ELIMINATION OF DENSITY-DRIVEN SEPARATION

Density-driven separation occurs when alloy constituents of different densities segregate under the influence of gravity during cooling and solidification. This phenomenon leads to:

- Non-uniform mechanical properties: Different regions of the alloy exhibit varying strength, toughness, and fatigue resistance.
- Localized weaknesses: Areas with higher concentrations of certain elements may become more prone to fracture or corrosion.
- Performance inconsistencies: Functional properties, such as electrical conductivity or thermal resistance, vary across the alloy.

In microgravity, density-driven separation is essentially eliminated, as gravitational forces are too weak to drive the separation of alloy constituents. This allows:

- Homogeneous Phase Distributions: All alloying elements remain uniformly mixed throughout the solidification process.
- Consistent Grain Structure: The absence of gravitational convection ensures even crystal nucleation and growth.
- Enhanced Mechanical Integrity: The resulting alloys are free from density-related weak spots and perform predictably under stress and extreme temperatures.

Microreactors play a pivotal role in preventing density separation by ensuring controlled precursor delivery, laminar flow mixing, and uniform cooling rates, resulting in alloys with consistent microstructures and reliable performance.

The synthesis of HEAs, SMAs, and lightweight metallic composites in microgravity represents a revolutionary advancement in metallurgical engineering and material science. The elimination of density-driven separation, combined with improved homogeneity and precise control over alloy microstructures, enables the creation of next-generation materials tailored for extreme environments and critical applications in space exploration and terrestrial industries. With the integration of microreactor technology, scientists and engineers gain an unprecedented level of control over alloy synthesis, cooling rates, and phase transformations, paving the way for materials that are lighter, stronger, more durable, and functionally superior. As humanity ventures deeper into space, these advanced alloys will play a vital role in ensuring the reliability, sustainability, and efficiency of future missions and technologies.

Applications of Metal Alloys Synthesized in Microgravity

The synthesis of advanced metal alloys in microgravity, including HEAs, SMAs, and lightweight metallic composites, has profound implications for the structural integrity and durability of spacecraft components and the corrosion resistance of materials used in long-term missions. In the unforgiving conditions of space, marked by extreme thermal fluctuations, cosmic radiation, micrometeoroid impacts, and chemically reactive environments, materials must exhibit exceptional mechanical strength, structural reliability, and long-term durability. Microgravity facilitates the creation of defect-free, homogeneous alloys with optimized microstructures, enabling advancements that are critical for the success and safety of space missions.

Structural Components for Spacecraft

Structural components of spacecraft, including hulls, support frames, propulsion system housings, fuel storage tanks, and landing gear, must meet stringent requirements for strength-to-weight ratio, fatigue resistance, and stability under dynamic loads. The materials used in these components must be lightweight to minimize launch costs while maintaining the mechanical properties required to withstand vibrational stress during launch, micro-meteoroid impacts, and prolonged exposure to vacuum and temperature extremes.

Lightweight Metallic Composites for Load-Bearing Structures

Lightweight alloys, such as aluminum-lithium (Al-Li) composites, Ti-Al alloys, and magnesium-based alloys, are indispensable for spacecraft construction due to their low density, high tensile strength, and excellent fatigue resistance. On Earth, synthesizing these alloys often results in uneven particle distribution, void formation, and grain boundary defects caused by gravitational forces. These imperfections compromise the alloy's load-bearing capacity and resistance to thermal expansion.

- Microgravity Advantage: In microgravity, density-driven separation is eliminated, ensuring a homogeneous distribution of alloying elements across the metallic matrix. This uniformity enhances the mechanical integrity, fatigue resistance, and damage tolerance of lightweight composites.
- Application Example: Microgravity-synthesized Al-Li alloys are used for spacecraft fuselage and structural panels, where lightweight yet durable materials are critical for reducing overall mass without sacrificing strength.

High-Entropy Alloys for Extreme Conditions

High-entropy alloys (HEAs) are particularly well-suited for structural applications in space due to their multi-element composition, high mechanical strength, and excellent thermal stability. These alloys are resistant to thermal creep, making them ideal for use in engine nozzles, re-entry shields, and TPSs.

- Microgravity Advantage: In space-based synthesis, the absence of gravitational convection ensures consistent elemental distribution and prevents localized phase separations that degrade alloy performance.
- Application Example: Space-grown HEAs are used in rocket engine components and spacecraft shielding systems, where materials must endure intense heat and mechanical stresses during launch and re-entry.

Shape-Memory Alloys for Adaptive Structures

Shape-memory alloys (SMAs), such as NiTi alloys, are capable of recovering their original shape after deformation when exposed to specific temperatures. These alloys are essential for deployable spacecraft structures, self-repairing panels, and adaptive mechanisms.

- Microgravity Advantage: Microgravity synthesis allows for precise phase transformation control, minimizing defects in the alloy microstructure and ensuring consistent actuation behavior.
- Application Example: NiTi-based SMAs are used in deployable solar panels, self-actuating hinges, and thermal control surfaces on spacecraft.

Structural Reliability for Long-Duration Missions

In long-term space missions, structural components are subjected to prolonged exposure to vacuum, temperature cycling, and mechanical stresses. Microgravity-synthesized alloys offer enhanced microstructural stability and resistance to fatigue cracking, ensuring reliable performance over extended mission durations.

- Application Example: Ti-Al alloys produced in microgravity are utilized in landing gear systems and structural trusses for spacecraft, where resistance to stress-induced fatigue and failure is critical.

The synthesis of structural alloys in microgravity enables the development of lighter, stronger, and more reliable materials, ensuring the safety, efficiency, and durability of spacecraft during launch, operation, and re-entry.

Corrosion-Resistant Materials for Long-Term Missions

Corrosion is one of the greatest threats to spacecraft durability and reliability, particularly during long-duration missions to Mars, lunar bases, or deep space exploration. In space, corrosion mechanisms differ from those on Earth, with contributing factors including:

- Atomic oxygen (AO) erosion in LEO.
- Radiation-induced chemical reactions.

- Electrochemical corrosion in spacecraft cooling and life-support systems.
- Moisture and biological contaminants in closed-loop habitats.

The development of corrosion-resistant alloys and coatings is therefore crucial for space habitat longevity, environmental control systems, and fuel storage tanks.

High-Entropy Alloys (HEAs) for Corrosion Resistance

HEAs are renowned for their exceptional resistance to corrosion across a wide range of temperatures and chemical environments. Their multi-element composition creates a self-healing passive oxide layer that protects the material from aggressive agents such as atomic oxygen and chlorine-based compounds.

- Microgravity Advantage: Microgravity enables the uniform distribution of alloying elements, ensuring a consistent passive oxide layer across the surface and preventing localized corrosion pits.
- Application Example: Space-grown HEAs are used for external spacecraft structures, orbital platforms, and protective casings for sensors and communication equipment in LEO, where atomic oxygen corrosion is a significant concern.

Lightweight Metallic Composites for Fuel and Water Storage Tanks

Fuel storage tanks and water reservoirs in spacecraft must endure pressurized environments, thermal cycling, and exposure to chemically reactive substances. Magnesium and titanium-based metallic composites are widely used due to their low density, mechanical strength, and resistance to chemical reactions.

- Microgravity Advantage: In microgravity, composites exhibit uniform microstructures and optimized particle reinforcement, reducing the risk of stress corrosion cracking and chemical degradation.
- Application Example: Ti-Al alloys synthesized in microgravity are used for cryogenic fuel tanks, offering improved resistance to hydrogen embrittlement and liquid oxygen exposure.

Shape-Memory Alloys for Corrosion-Sensitive Systems

Shape-memory alloys, particularly NiTi alloys, are highly resistant to corrosion in aqueous and chemically aggressive environments. Their self-healing and adaptive properties make them valuable for space coolant loops, life-support plumbing, and adaptive sealing mechanisms.

- Microgravity Advantage: Microgravity ensures uniform alloy microstructures and consistent passive oxide layers, enhancing their resistance to chemical degradation and reducing the likelihood of microcrack formation.
- Application Example: NiTi alloys are used in self-sealing valves and corrosion-resistant actuators for life-support and fluid transport systems in spacecraft.

Protective Coatings for Spacecraft Exteriors

In addition to bulk alloy synthesis, microgravity enables the creation of advanced ceramic-metallic (cermet) coatings on alloy substrates. These coatings offer superior resistance to oxidation, radiation damage, and chemical erosion.

- Microgravity Advantage: Coatings applied in microgravity exhibit uniform layer thickness, minimal porosity, and enhanced adhesion to the alloy substrate.
- Application Example: TBCs synthesized in microgravity are applied to rocket engine nozzles and spacecraft thermal control panels, extending their operational lifespan.

The synthesis of metal alloys in microgravity enables the development of materials with unprecedented homogeneity, corrosion resistance, and mechanical integrity. Structural alloys contribute to lighter, more durable spacecraft components, while corrosion-resistant materials ensure long-term reliability and environmental resilience during extended missions. Whether it's fuel storage tanks, deployable structures, or spacecraft hulls, these advanced alloys address the harsh challenges of space environments while driving innovation in aerospace engineering and long-term space colonization. As space missions become increasingly ambitious, these microgravity-synthesized alloys will form the backbone of sustainable, reliable, and efficient space systems.

SYNTHESIS OF POLYMERS IN MICROGRAVITY

The synthesis of polymers in microgravity represents a transformative advancement in material science, enabling the production of conductive polymers, high-strength elastomers, and biocompatible polymers with unprecedented uniformity, structural integrity, and functional performance. Freed from gravitational effects such as sedimentation, phase separation, and thermal convection, microgravity environments allow diffusion-driven polymerization and precise molecular alignment, resulting in materials with consistent chain growth, minimal defects, and tailored properties. With the integration of microreactor technology, scientists gain real-time control over polymerization conditions, enabling scalable synthesis of polymers optimized for extreme environments and critical applications in space exploration.

These high-performance polymers are central to the development of next-generation systems for spacecraft and space habitats. Conductive polymers support lightweight electronics and energy systems; elastomers provide durable seals and shock-absorbing structures; and biocompatible polymers enable personalized medical devices, drug delivery systems, and tissue engineering tools for long-duration missions. Additionally, space-synthesized polymers are revolutionizing insulation materials by enhancing thermal control and radiation shielding, while enabling the on-demand production of biomedical tools through 3D printing. As humanity expands its presence in space, microgravity-enabled polymer synthesis will be essential to building resilient, adaptable, and self-sustaining technologies for long-term exploration and habitation beyond Earth.

The synthesis of polymers in microgravity represents a significant leap in material science, offering unique advantages for producing conductive polymers, high-strength elastomers, and biocompatible polymers. These classes of polymers are fundamental in applications ranging from space electronics and structural components to biomedical devices and life-support systems. On Earth, polymer synthesis often suffers from gravitational effects such as sedimentation, phase separation, and uneven thermal gradients, which can introduce structural defects, inconsistent chain growth, and irregular material properties. In microgravity, these challenges are largely eliminated, creating an ideal environment for the precise control of polymerization reactions and the synthesis of materials with uniform structures and superior performance characteristics.

Microgravity provides an environment where diffusion-driven processes dominate over convection and sedimentation, allowing researchers to achieve unprecedented control over monomer distribution, chain propagation, and molecular alignment during polymerization. Coupled with the precision of microreactor technology, scientists can fine-tune reaction conditions such as temperature, pressure, monomer concentration, and catalyst activation, leading to polymers with minimal defects, consistent chain lengths, and tailored properties for highly specialized applications.

CONDUCTIVE POLYMERS

Conductive polymers, such as polyaniline (PANI), polypyrrole (PPy), and poly(3,4-ethylenedioxythiophene) (PEDOT), have revolutionized the fields of flexible electronics, energy storage, and sensor technologies. These materials combine the electrical conductivity of metals with the flexibility and lightweight characteristics of polymers, making them ideal for use in wearable sensors, flexible circuits, and spacecraft energy systems.

On Earth, the synthesis of conductive polymers is often limited by inhomogeneous monomer distribution, uneven catalyst activity, and thermal gradients, which result in irregular polymer chain growth, reduced conductivity, and inconsistent mechanical properties.

- Microgravity Advantage: In microgravity, monomer dispersion and catalyst activation become highly uniform, allowing polymer chains to grow consistently without defects or irregularities. This ensures that the resulting conductive polymers exhibit higher charge carrier mobility, enhanced conductivity, and greater mechanical flexibility.
- Application Example: Space-grown conductive polymers can be used in flexible solar panels, lightweight batteries, and embedded electronic circuits on spacecraft, where weight and energy efficiency are critical constraints.

Furthermore, microreactor technology allows precise control over polymer doping and chain alignment, enabling scientists to tailor electronic and optical properties for specific space applications, such as radiation-resistant coatings and thermoelectric devices.

High-Strength Elastomers

High-strength elastomers are polymers characterized by their elasticity, durability, and resistance to extreme mechanical stress. These materials are widely used in seals, gaskets, shock absorbers, and flexible connectors in spacecraft systems. Elastomers such as silicone rubber, polyurethane, and polychloroprene are indispensable for dynamic sealing, vibration dampening, and thermal insulation.

On Earth, synthesizing elastomers with consistent mechanical properties is challenging due to density gradients, phase separation, and inconsistent cross-linking reactions driven by gravity. These factors often result in uneven material properties, weak spots, and poor stress distribution.

- Microgravity Advantage: In microgravity, cross-linking reactions and molecular chain alignment occur uniformly, resulting in elastomers with consistent mechanical strength, superior elasticity, and improved stress tolerance across the material.
- Application Example: Space-grown elastomers can be utilized in flexible spacecraft joints, shock-absorbing mounts, and airtight seals in life-support systems, where long-term reliability and resistance to material fatigue are essential.

Additionally, microreactors allow for real-time monitoring and fine-tuned control of reaction kinetics, ensuring optimal cross-linking density and uniformity in polymer network structures. This results in high-performance elastomers capable of withstanding extreme thermal cycling and mechanical loads in space environments.

Biocompatible Polymers

Biocompatible polymers, such as PLA, polyethylene glycol (PEG), and PCL, are essential for medical devices, tissue scaffolding, and drug delivery systems. In space, these polymers have significant potential for on-demand medical manufacturing, personalized healthcare solutions, and advanced biological research aboard spacecraft.

On Earth, the synthesis of biocompatible polymers often suffers from impurity inclusions, non-uniform molecular weight distribution, and inconsistent degradation rates due to gravitational forces affecting polymerization dynamics and crystallization processes.

- Microgravity Advantage: In microgravity, monomer diffusion, polymer chain growth, and crystallization processes are uniform and predictable, allowing for the production of polymers with consistent molecular weights, fewer impurities, and highly controlled degradation profiles.
- Application Example: Space-grown biocompatible polymers can be used in 3D-printed surgical implants, biodegradable wound dressings, and controlled-release drug delivery capsules tailored to astronauts' specific medical needs.

Microreactors provide the precision required to control monomer-to-polymer conversion rates, molecular weight distribution, and chain architecture, ensuring biocompatibility, stability, and performance predictability of the synthesized polymers. This capability enables personalized medicine and medical device manufacturing on long-duration space missions, reducing reliance on Earth-based supply chains.

Polymerization Control in Zero Gravity

The success of polymer synthesis in microgravity depends heavily on precise control over polymerization reactions, including chain growth, molecular weight distribution, and defect prevention. Polymerization is a complex chemical process influenced by factors such as monomer concentration, reaction temperature, and catalyst activity, all of which must be finely balanced to achieve optimal polymer properties.

Uniform Polymer Chain Growth

In terrestrial environments, gravitational effects often lead to density-driven convection and phase separation, causing uneven monomer distribution during polymerization. These inconsistencies result in variable chain lengths, incomplete polymerization, and structural defects within the final material.

- Microgravity Advantage: In microgravity, diffusion dominates over convection, ensuring uniform distribution of monomers and catalysts throughout the reaction medium. This leads to consistent chain growth and reduced variability in molecular weights.
- Application Example: Conductive polymers synthesized in microgravity exhibit longer, more uniform polymer chains, enhancing their electrical conductivity and mechanical flexibility.

Minimal Defect Formation

Defects such as voids, entanglements, and cross-linking irregularities often occur during polymer synthesis on Earth due to thermal gradients and gravitational convection. These defects can weaken the mechanical properties and reduce the functional performance of the polymer.

- Microgravity Advantage: In microgravity, thermal gradients are minimized, and uniform heat distribution ensures consistent reaction rates across the polymerizing medium. This eliminates the formation of weak spots and irregular cross-linking patterns in elastomers and biocompatible polymers.
- Application Example: Biocompatible polymers synthesized in microgravity display enhanced structural integrity, predictable degradation profiles, and improved safety for biomedical implants and drug delivery systems.

Real-Time Monitoring and Adjustments via Microreactors

Microreactors play a critical role in enabling real-time monitoring and dynamic control of polymerization reactions in microgravity. Integrated sensors track temperature, pressure, viscosity, and reaction kinetics, allowing scientists to make on-the-fly adjustments to optimize chain growth and prevent defects.

- Enhanced Precision: Microreactors allow for precise catalyst activation, reactant flow rates, and temperature control, ensuring uniform polymer chain growth.
- Scalability: The modular nature of microreactors allows the production of polymers in both small experimental batches and larger commercial scales in space-based manufacturing facilities.

The synthesis of conductive polymers, high-strength elastomers, and biocompatible polymers in microgravity, supported by advanced microreactor technology, represents a paradigm shift in material science and space-based manufacturing. The unique advantages of microgravity, including uniform chain growth, reduced defect formation, and precise reaction control, enable the creation of high-performance polymers with exceptional functional properties. These materials will play a pivotal role in spacecraft electronics, structural components, medical systems, and advanced biological research. As humanity pushes the boundaries of deep-space exploration and space colonization, microgravity-synthesized polymers will remain the cornerstones of sustainable and innovative technological solutions.

Applications of Polymers Synthesized in Microgravity

The synthesis of polymers in microgravity, including conductive polymers, high-strength elastomers, and biocompatible polymers, offers unique opportunities to create materials with unparalleled structural integrity, functional precision, and application-specific properties. These materials are essential for addressing critical challenges in space habitats, where reliability, sustainability, and resource efficiency are paramount. Two key application areas stand out: advanced insulation materials for spacecraft and habitats, and biomedical devices tailored for long-duration missions.

Advanced Insulation Materials

In the extreme and unpredictable environment of space, thermal insulation is a fundamental requirement for spacecraft, space habitats, and scientific equipment. Spacecraft and habitats are exposed to severe temperature fluctuations ranging from –270°C in the cold vacuum of space to over 200°C in direct sunlight. Additionally, insulation must address radiation shielding, mechanical durability, and low outgassing characteristics to ensure long-term functionality and safety.

Thermal Insulation Polymers

Polymers such as aerogels, polyimides (e.g., Kapton), and silicone-based foams are widely used for thermal insulation due to their low thermal conductivity, lightweight properties, and high durability. However, producing these materials on Earth is challenging due to density-driven gradients, inconsistent cross-linking, and microstructural defects caused by gravitational forces.

- Microgravity Advantage: In microgravity, polymers can be synthesized with uniform pore structures, consistent cross-linking densities, and minimal defects, resulting in superior thermal insulation performance. The absence of convection ensures that the polymerization process proceeds uniformly, creating materials with highly consistent thermal properties across their structure.

- Application Example: Microgravity-synthesized polyimide films and aerogels can be used as insulating layers for spacecraft hulls, solar panels, and cryogenic fuel storage tanks.

Radiation-Resistant Polymers

Space habitats and spacecraft must also be shielded from cosmic radiation and solar particle events. Certain polymers, such as polyethylene-based materials and hydrogen-rich polymeric compounds, are effective at absorbing and attenuating radiation due to their high hydrogen content and molecular structure.

- Microgravity Advantage: In microgravity, polymer chains align uniformly, creating a consistent molecular network that enhances radiation absorption efficiency while maintaining mechanical flexibility.
- Application Example: Hydrogen-rich polymers synthesized in microgravity can serve as radiation shielding layers for crew cabins, sleeping quarters, and storage compartments in space habitats.

Multifunctional Insulating Polymers

Future insulation materials will need to integrate multiple functions, such as thermal control, radiation shielding, and mechanical resilience. Polymers synthesized in microgravity can be engineered to have hybrid properties, combining features such as self-healing capabilities, adaptability to thermal cycling, and embedded sensors for real-time performance monitoring.

- Application Example: Space-grown self-healing polymers can be integrated into flexible insulation blankets used in habitats, automatically sealing microcracks caused by thermal stress or micrometeoroid impacts.

In essence, microgravity-synthesized polymers are paving the way for next-generation thermal and radiation insulation materials that are lighter, more efficient, and longer-lasting, ensuring the safety and sustainability of spacecraft and space habitats.

Biomedical Devices for Space Habitats

Long-duration space missions, such as those to Mars, lunar bases, or orbital habitats, present unique biomedical challenges. Astronauts face limited access to medical supplies, prolonged exposure to microgravity, and physiological changes, including bone density loss, muscle atrophy, and altered immune responses. Biocompatible polymers synthesized in microgravity offer transformative solutions for developing advanced medical devices, implants, and drug delivery systems tailored to these conditions.

Biocompatible Polymers for Implants and Prosthetics

Biocompatible polymers, such as PLA, PEG, and PCL, are widely used in surgical implants, prosthetic devices, and tissue scaffolding. These polymers must exhibit predictable degradation rates, mechanical stability, and compatibility with human tissues.

- Microgravity Advantage: In microgravity, polymerization occurs under uniform diffusion and thermal conditions, ensuring consistent molecular weight distribution, minimal impurities, and optimized mechanical properties. This level of control results in implants with better biointegration, reduced inflammatory responses, and longer functional lifespans.
- Application Example: Space-grown PLA scaffolds can be used for 3D-printed bone implants and regenerative tissue scaffolding to address bone loss caused by prolonged microgravity exposure.

Drug Delivery Systems

Controlled-release drug delivery systems are essential in space missions, where medical intervention capabilities are limited. Polymers such as poly(lactic-co-glycolic acid) (PLGA) and hydrogels are commonly used in drug delivery due to their biocompatibility, controlled degradation rates, and ability to encapsulate pharmaceuticals.

- Microgravity Advantage: In microgravity, drug-loaded polymer matrices can be synthesized with uniform encapsulation and consistent porosity, ensuring precise and predictable drug release rates. This is particularly valuable for long-term therapies required for conditions like osteoporosis, chronic pain, or immune system modulation.
- Application Example: Space-grown polymeric drug delivery capsules can provide sustained medication release to astronauts, reducing the need for frequent medical intervention.

Wearable Medical Sensors

Conductive polymers, such as PPy and PANI, enable the creation of wearable biosensors and health-monitoring devices. These materials are flexible, lightweight, and capable of real-time health data acquisition and wireless transmission.

- Microgravity Advantage: In microgravity, conductive polymers exhibit consistent chain alignment and minimal surface defects, improving signal sensitivity, conductivity, and mechanical resilience.
- Application Example: Flexible health-monitoring sensors embedded in space suits can track vital signs, hydration levels, and muscle activity, enabling real-time health assessments during extravehicular activities (EVAs).

Self-Sterilizing and Antimicrobial Polymers

Microgravity environments create unique challenges for microbial growth and contamination in enclosed habitats. Polymers with antimicrobial and self-sterilizing properties are critical for maintaining clean and safe living conditions.

- Microgravity Advantage: In microgravity, polymer synthesis allows even distribution of antimicrobial agents, such as silver nanoparticles or iodine compounds, across the polymer matrix.
- Application Example: Antimicrobial polymer coatings synthesized in space can be applied to water purification systems, surgical instruments, and high-contact surfaces in space habitats.

3D Printing with Space-Synthesized Polymers

3D printing, or additive manufacturing, is becoming a cornerstone of space missions for producing on-demand tools, medical implants, and replacement parts. Biocompatible polymers synthesized in microgravity are ideal feedstocks for 3D printing medical devices due to their purity, consistent viscosity, and predictable mechanical behavior.

- Application Example: Space habitats equipped with 3D bioprinters using microgravity-synthesized polymers can produce custom prosthetics, surgical instruments, and tissue scaffolds tailored to individual astronaut needs.

The synthesis of polymers in microgravity, including conductive polymers, high-strength elastomers, and biocompatible polymers, holds enormous potential for addressing critical needs in space habitats and long-term missions. Advanced insulation materials derived from space-synthesized polymers offer enhanced thermal control, radiation shielding, and multifunctionality, ensuring habitat safety and efficiency. Similarly, biocompatible polymers enable the development of personalized medical devices, drug delivery systems, and health-monitoring technologies, improving astronaut healthcare and resilience during extended space missions.

As space agencies and private enterprises continue to expand humanity's presence beyond Earth, microgravity-synthesized polymers will play a central role in creating self-sustaining, adaptable, and resource-efficient space habitats, ultimately supporting humanity's ambitions for sustainable living and exploration on the Moon, Mars, and beyond.

CHAPTER 5

Integration with Space Research Platforms

The integration of microreactors into space research platforms represents a transformative step forward in in-situ material synthesis, resource utilization, and scientific experimentation. These advanced systems are compact, energy-efficient, and capable of conducting complex chemical and material synthesis processes with high precision and minimal human intervention. The International Space Station (ISS) serves as a critical testbed for microreactor technologies, while autonomous microreactor systems pave the way for long-duration deep-space missions. Furthermore, the ability to collect, monitor, and transmit data in real time ensures that ground-based scientists can oversee and optimize space-based chemical processes. These platforms not only advance scientific research but also establish the foundation for self-sustaining, efficient, and scalable space exploration systems.

INTEGRATION OF MICROREACTOR SYSTEMS WITH SPACE RESEARCH PLATFORMS

The integration of microreactor systems into space research platforms has emerged as a transformative leap in enabling ISRU, material synthesis, and sustainable operations in space. These compact, high-precision systems are designed to perform chemical and material processing under the unique conditions of microgravity, offering a path toward greater

DOI: 10.1201/9781003656081-5

autonomy, efficiency, and adaptability for long-term space missions. Central to the advancement of microreactor technology is its incorporation into the ISS, its evolution into autonomous systems for deep-space missions, and the growing use of real-time data collection to monitor and optimize performance from Earth.

On the ISS, microreactors have become essential tools for chemical and pharmaceutical research. Integrated into dedicated laboratory modules such as the Microgravity Science Glovebox (MSG) and the Life Sciences Glovebox (LSG), these reactors benefit from the station's advanced environmental controls, safety infrastructure, and experimental versatility. Their compact, modular design allows multiple units to operate in parallel, making efficient use of the ISS's limited space while supporting scalable experimentation. Safety remains paramount; integrated chemical containment, pressure management systems, thermal regulation, and automated shutdown protocols ensure the safe execution of sensitive processes. Energy efficiency is also a key advantage, as these systems are optimized to consume minimal power while maintaining high productivity levels.

Microreactor experiments aboard the ISS are typically conducted under a combination of astronaut support and ground-based remote operation. Astronauts handle setup, sampling, and maintenance, while scientists on Earth remotely monitor and control reactor functions using telemetry systems. Integrated sensors ranging from spectrometers and thermal monitors to pressure and imaging systems enable in-situ synthesis and analysis, providing a real-time view of chemical transformations and process conditions. Synthesized materials are often preserved and returned to Earth for further characterization using advanced instrumentation. Applications range from the synthesis of high-purity crystals, polymers, and alloys to the on-demand production of pharmaceuticals and the study of chemical reaction kinetics and phase transitions in a gravity-free environment. The ISS thus serves as both a proving ground and operational hub for space-based microreactor technologies.

Beyond LEO, autonomous microreactor systems represent a critical step toward supporting deep-space missions to the Moon, Mars, and beyond. These systems are engineered for continuous, self-sustaining operation with minimal human intervention, a necessity given the long durations, isolation, and harsh conditions of such missions. Designed to withstand cosmic radiation, temperature extremes, and micrometeoroid impacts, autonomous microreactors rely on advanced AI and machine learning (ML) algorithms to monitor and adjust reaction parameters in real time. They often function as part of a closed-loop system, recycling waste products and minimizing the need for external inputs.

One of the most promising aspects of autonomous microreactors is their ability to support in-situ manufacturing and ISRU. For example, they can convert carbon dioxide from the Martian atmosphere into oxygen and methane fuel, purify water extracted from lunar ice deposits, or process locally sourced materials into metals, ceramics, and polymers for habitat construction and equipment maintenance. Their modular nature allows for scalable deployment: multiple units can be operated in parallel to increase output, with redundancy built into the system to ensure continued function even if one unit fails. AI-driven controls enable dynamic resource allocation and synthesis prioritization based on mission demands. These capabilities make autonomous microreactors essential for building self-sufficient extraterrestrial colonies, enabling human missions that do not rely on continuous resupply from Earth.

The ability to monitor and optimize microreactor operations in real time is enabled by advanced data integration and sensor technologies. Modern microreactors are equipped with temperature and pressure sensors, spectroscopic analyzers, and flow meters that continuously measure critical reaction parameters. This sensor data is transmitted back to Earth using high-bandwidth communication systems, allowing ground-based scientists to track experiments, adjust variables, and troubleshoot issues as they arise. Real-time data not only supports operational efficiency but also provides valuable insights into chemical processes under microgravity conditions, advancing our understanding of space-based material synthesis.

AI plays a crucial role in managing these complex data streams. AI platforms process vast quantities of sensor data to predict reactor performance, identify anomalies, and fine-tune operating conditions. These systems can also make autonomous decisions when communication delays make real-time human intervention impractical. For example, in deep-space environments where communication latency may reach several minutes, AI algorithms can detect suboptimal reaction conditions and adjust flow rates or temperatures without waiting for commands from Earth. Applications of this integration include remote diagnostics, automated safety compliance, process optimization, and the advancement of scientific discovery through continuous experimentation.

CHALLENGES AND FUTURE DIRECTIONS IN SPACE-BASED MICROREACTOR SYSTEMS

The integration of microreactor systems into space exploration platforms marks a critical advancement in achieving ISRU, enabling the synthesis of advanced materials and supporting sustainable operations in extraterrestrial environments. These compact and precise chemical reactors are designed to function in microgravity and resource-limited conditions, offering modularity, high efficiency, and low energy consumption. However, the deployment of microreactors in space is not without its challenges. Several technical, operational, and collaborative barriers must be addressed to fully realize their potential in long-duration space missions and planetary habitats. Future development must focus on overcoming reactor clogging, improving maintenance strategies, addressing scalability constraints, and fostering collaborative innovation through international and public–private partnerships.

One of the most pressing technical challenges in space-based microreactor systems is reactor clogging. On Earth, such issues are typically mitigated through routine maintenance, disassembly, or flushing protocols. In space, where microgravity alters fluid dynamics and limits sedimentation, solid particles whether precipitates, polymer residues, or crystallized reagents can remain suspended within reactor fluids and gradually accumulate inside microchannels. Over time, these deposits block fluid flow, reduce reaction efficiency, and may cause dangerous pressure build-ups, potentially leading to system failure. Addressing clogging in microgravity requires a combination of passive and active solutions. Self-cleaning designs using ultrasonic agitation or piezoelectric vibrations can help dislodge deposits before they cause obstructions. Additionally, adaptive flow control systems, guided by real-time sensors, can identify early signs of channel blockage and adjust fluid velocity or pressure to prevent buildup. In critical scenarios, modular microreactor designs would allow astronauts to easily replace clogged reactor units without the need for full system disassembly, streamlining repair and minimizing downtime.

Maintenance in space presents another set of challenges. Microreactor systems often contain delicate, miniaturized components that are difficult to access and manipulate under the constrained conditions of a spacecraft or habitat. Repairs demand both specialized tools and extensive astronaut training, complicating mission planning and increasing operational risk. To address this, next-generation microreactors must incorporate advanced diagnostic tools capable of detecting faults autonomously and reporting their status through integrated telemetry systems. Remote operation and troubleshooting by ground-based teams will be crucial, allowing Earth-based experts to assess problems using real-time data and propose corrective measures without requiring immediate astronaut intervention. Looking further ahead, autonomous repair technologies such as reactors with replaceable microfluidic modules or components built from self-healing materials could provide robust solutions to the maintenance challenge, ensuring the longevity and resilience of microreactor systems during multiyear missions.

Scalability represents a third major hurdle. While microreactors are exceptionally efficient for small-scale chemical synthesis, scaling them to meet the demands of long-duration missions, resource processing, or industrial-grade material production remains a challenge. Unlike conventional reactors, which can simply be enlarged, microreactor systems must be scaled horizontally by operating numerous units in parallel. This approach increases the complexity of system integration, especially in terms of coordinating control protocols, resource allocation, and energy distribution. Power and cooling requirements also scale with the number of active units, imposing additional strain on spacecraft infrastructure. To address these constraints, microreactors can be deployed in modular arrays managed by centralized control units and AI-driven algorithms capable of dynamic load balancing. These systems would optimize energy usage, prioritize mission-critical synthesis, and reallocate resources based on system performance and real-time needs. Innovations in energy efficiency such as incorporating space-based solar power and low-energy chemical pathways will also be key to making large-scale microreactor deployment viable.

Overcoming these technical challenges clogging, maintenance, and scalability requires a multidisciplinary approach that blends chemical engineering, materials science, automation, AI, and space systems design. Future research and development should focus on enhancing reactor reliability, implementing predictive maintenance, and creating scalable, self-regulating systems that can operate with minimal human oversight. These innovations will be essential not only for enabling microreactor functionality in space but also for supporting broader goals of sustainable habitation and industrialization beyond Earth. As space missions become more ambitious and long-term, solving these challenges will be critical to transforming microreactor technologies from experimental tools into indispensable infrastructure for off-world living and exploration.

LONG-TERM SUSTAINABILITY AND COLLABORATIVE INNOVATION IN SPACE-BASED CHEMICAL SYSTEMS

The long-term success of space missions hinges on the development of a circular chemical economy, where reagents, solvents, and by-products are continually recycled, purified, and reused. This approach is essential not only for minimizing dependence on resupply missions from Earth but also for establishing sustainable space habitats that can function autonomously. In environments where logistical support is limited and resources are scarce, efficient waste management and resource recovery systems are vital to reduce environmental impact and maintain operational continuity.

Reagent recycling is a cornerstone of chemical sustainability in space. During synthesis processes carried out within microreactors, reagents are often consumed or degraded, which traditionally would necessitate periodic resupply. However, the feasibility of long-term missions depends on the ability to recover and reuse these substances. Recycling in space presents several challenges, including the complexity of chemical waste streams, the energy requirements of separation and purification, and the gradual degradation of reagents over multiple cycles. Addressing these challenges requires the integration of closed-loop chemical systems capable of real-time reagent recovery, as well as the strategic use of ISRU techniques that harness locally available materials such as Martian carbon dioxide or lunar regolith to replenish chemical inventories. Moreover, AI-optimized systems are emerging as a critical component of space recycling, employing ML algorithms to dynamically adjust reaction parameters and maximize recovery efficiency while minimizing waste.

Alongside reagent recycling, waste management is equally critical to maintaining a clean, habitable, and functional space environment. Without proper waste handling, chemical and biological by-products can compromise air quality, corrode equipment, or pose direct threats to crew health. Space habitats, with their limited storage and isolation, cannot afford traditional waste disposal approaches. Complex waste streams arising from microreactor operations further complicate disposal strategies, often containing toxic, reactive, or volatile compounds. To mitigate these issues, new solutions are being developed, such as on-demand waste recycling, which converts waste into usable materials or energy sources;

solidification techniques that immobilize hazardous substances in stable matrices to prevent leakage; and automated segregation systems that classify and process waste based on real-time sensor data. These approaches, particularly when integrated with AI-driven control systems, provide the foundation for a sustainable closed-loop life support and chemical processing ecosystem in space.

The advancement of such sophisticated systems depends heavily on collaboration between international space agencies, research institutions, and private aerospace companies. Effective partnerships are essential for sharing resources, accelerating innovation, and standardizing practices across global missions. Agencies like NASA, ESA, Roscosmos, and JAXA have increasingly embraced the benefits of collaborative research, combining infrastructure such as the ISS and potential lunar or Martian outposts to reduce individual mission costs and increase the scope of experimental research. Joint research initiatives in areas like microgravity synthesis, pharmaceutical development, and chemical recycling enable cross-validation of results and expand the collective knowledge base, while the establishment of harmonized safety and operational standards facilitates seamless integration of microreactors into multinational missions.

Private companies, including SpaceX, Blue Origin, and Sierra Space, are also becoming key players in this collaborative ecosystem. Their capacity for rapid prototyping, innovation, and commercial deployment complements the more methodical approach of public space agencies. As commercial interest in space-based manufacturing grows, the production of high-value materials in microgravity such as pharmaceuticals, polymers, and alloys offers new economic opportunities. Public–private partnerships align the goals of science and commerce, enabling the development of space technologies that are both scientifically robust and economically viable.

Academic institutions and research organizations play a pivotal role in this ecosystem as well, offering expertise in reactor design, chemical engineering, and sustainable process development. Their contributions extend to knowledge-sharing platforms, where data, methodologies, and findings are disseminated openly to foster global collaboration and accelerate progress. By building integrated research and development ecosystems that span academia, industry, and government, the space community can more effectively overcome technical barriers and develop the infrastructure needed for sustainable off-world habitation.

CHAPTER 6

Going Forward

The integration of microreactor systems into space exploration marks a transformative step in humanity's journey beyond Earth, enabling the precise synthesis of next-generation materials, pharmaceuticals, and fuels with properties unattainable under terrestrial conditions. These compact, modular systems offer unparalleled control over chemical reactions, leveraging the unique environment of microgravity to produce defect-free crystals, high-performance alloys, and advanced polymers. Their use on platforms such as the International Space Station and future lunar and Martian outposts supports a new frontier in space-based manufacturing, where autonomous microreactors convert local resources into essential compounds, paving the way for sustainable, long-duration missions and permanent extraterrestrial settlements.

To fully realize this vision, continued investment in microreactor technology is essential. Addressing challenges such as reactor clogging, scalability, and closed-loop sustainability will require advancements in AI-driven monitoring, self-healing materials, and in-situ resource utilization. Furthermore, success hinges on collaborative efforts between governments, private industry, and academia to establish standards, share knowledge, and commercialize the breakthroughs of space-based production. As we shift from exploring space to inhabiting and building within it, microreactors will become the backbone of a new space economy one that fuels discovery, enhances life on Earth, and enables humanity to thrive in the cosmos.

DOI: 10.1201/9781003656081-6

TRANSFORMATIVE POTENTIAL OF MICROREACTORS IN SPACE RESEARCH

Enabling Next-Generation Materials with Space-Specific Properties

The integration of microreactor systems into space research platforms represents a transformative leap in our ability to synthesize advanced materials, pharmaceuticals, and chemical compounds in microgravity and extraterrestrial environments. Unlike traditional terrestrial manufacturing methods, microreactors offer a highly controlled, efficient, and scalable platform for material synthesis, enabling the creation of next-generation materials with unique properties unattainable under Earth's gravitational constraints. Whether it's the precision synthesis of semiconductor crystals for optical systems, the production of heat-resistant ceramics for spacecraft shielding, or the development of polymers for biomedical devices and flexible electronics, microreactors have demonstrated their ability to overcome longstanding limitations in material science.

In microgravity, the absence of convection currents, density-driven separation, and sedimentation allows researchers to achieve uniform particle distribution, defect-free crystallization, and consistent molecular structures across synthesized materials. This has direct implications for space exploration technologies, including more reliable communication systems, resilient spacecraft components, adaptive structures, and life-support systems. Beyond these immediate applications, microgravity-synthesized materials hold the potential to revolutionize Earth-based industries, from aerospace engineering and pharmaceuticals to sustainable manufacturing practices and environmental remediation.

Microreactors are not just tools for material synthesis; they are platforms for scientific discovery, engineering innovation, and mission sustainability. They empower researchers to explore unconventional chemical pathways, optimize reaction kinetics, and test novel hypotheses that were previously limited by terrestrial conditions. The knowledge gained from space-based synthesis will drive technological advancements on Earth and set the stage for humanity's long-term presence in space.

The Path Forward: Continued Investment and Innovation in Microreactor Technology for Space Environments

Microreactor systems have emerged as one of the most promising technological solutions for supporting chemical synthesis, resource utilization, and life support in space. Yet, despite their transformative potential, the full capabilities of microreactors remain underutilized. Unlocking this potential and transforming it into the foundation of sustainable human presence in space will require sustained investment, bold innovation, and coordinated collaboration across the scientific, governmental, and commercial spheres. The path forward is clear: we must strategically overcome technical limitations, build resilient systems, cultivate specialized expertise, and establish a robust framework for policy and cooperation.

Addressing the remaining technical challenges is central to advancing microreactor deployment in space. Reactor clogging, material degradation, scalability limitations, and maintenance constraints continue to pose operational risks in space missions. To mitigate these issues, the next generation of microreactors must incorporate advanced features such as self-healing materials, modular architectures for rapid replacement, and AI-driven diagnostics that enable real-time performance monitoring and predictive fault detection. Breakthroughs in materials science and fluid dynamics will further improve microreactor reliability, making them more suitable for the harsh and unpredictable conditions of space.

Equally important is the advancement of sustainable and circular chemical systems. Long-duration missions, including those to the Moon and Mars, will require microreactors to function as components of closed-loop systems where reagents, solvents, and by-products are continuously recycled. Developing robust processes for ISRU, such as extracting oxygen and water from lunar regolith or Martian ice, will further reduce reliance on Earth-based supplies. Integrated waste management and resource recovery systems will ensure that microreactor-based synthesis operations can operate indefinitely with minimal external input, laying the groundwork for self-sustaining habitats and industrial hubs beyond Earth.

Human factors will also play a critical role in the successful implementation of microreactor technologies. Astronauts and mission specialists must be equipped with specialized training in microreactor operations, maintenance procedures, and emergency response protocols. As microreactors become more integrated into life support, power generation, and manufacturing systems aboard spacecraft and planetary habitats, the ability to troubleshoot and adapt these technologies in real time will be vital to mission success and crew safety.

Cross-sector collaboration will be the driving force behind the next wave of microreactor innovation. Public space agencies, academic research institutions, private aerospace firms, and industry experts must work together to pool knowledge, share resources, and co-develop technologies. Joint research missions aboard the ISS and planned lunar platforms can serve as testbeds for experimental reactors, while open-source data-sharing initiatives will allow discoveries to be validated, scaled, and adapted more quickly. Partnerships between government and industry will also accelerate the commercialization of space-derived technologies for terrestrial use.

Establishing clear policies and standardized operational frameworks will be crucial to the widespread adoption of microreactor systems. A unified set of guidelines covering safety protocols, design parameters, integration procedures, and environmental impact assessments will ensure that microreactors can be seamlessly incorporated into diverse mission architectures, whether aboard orbital stations, planetary bases, or autonomous robotic outposts. International cooperation in defining these standards will also support interoperability across spacefaring nations and companies, creating a foundation for collaborative ventures on the Moon, Mars, and beyond.

Investing in microreactor technology is about more than enhancing the functionality of future space missions it is about creating a bridge between space and Earth. Many of the innovations developed for microgravity environments, such as closed-loop processing, autonomous manufacturing, and compact chemical reactors, will find powerful applications in terrestrial industries ranging from pharmaceuticals to clean energy. These technologies not only expand our reach into space but also improve life on Earth by driving forward scientific discovery, industrial efficiency, and environmental sustainability.

Vision for the Future: Space-Based Manufacturing Hubs Leveraging Microreactor Systems for Material Innovation

As humanity advances further into space, the future of exploration and colonization will be increasingly defined by our ability to produce essential resources independently of Earth. Central to this transformation is the development of space-based manufacturing hubs that harness microreactor systems to synthesize critical materials, medicines, and fuels on demand. These hubs strategically situated in LEO, on the Moon, and on Mars represent a paradigm shift in how we approach resource utilization, environmental sustainability, and technological innovation beyond our home planet.

In-orbit manufacturing platforms will serve as the first iteration of this future. Equipped with advanced microreactor modules, space stations will operate as orbital production centers capable of synthesizing high-performance alloys, defect-free crystals, and specialized polymers. These materials will be immediately available for spacecraft maintenance, scientific instrumentation, or the construction of new infrastructure in space. The microgravity environment offers distinct advantages in producing materials with enhanced properties that are difficult to replicate on Earth, positioning these platforms as crucial to both the functionality and longevity of space missions.

On extraterrestrial surfaces such as the Moon and Mars, microreactors will become integral components of ISRU systems. By converting locally available resources such as Martian carbon dioxide and lunar regolith into usable compounds like oxygen, water, methane fuel, and building materials, microreactors will drastically reduce dependence on Earth-based supply chains. These systems will not only enable astronauts to survive in remote environments but will also lay the groundwork for permanent settlements. The ability to produce essential resources onsite represents a foundational leap in establishing sustainable human operations in other worlds.

Moreover, the materials synthesized in space will not be limited to extraterrestrial use. Radiation-resistant alloys, high-purity optical crystals, and pharmaceutical precursors manufactured in microgravity will hold significant commercial value on Earth. As the economics of space transportation evolve, markets may emerge for space-produced semiconductors, medical implants, and advanced composites that outperform their terrestrial counterparts. In this way, microreactor-enabled production has the potential to drive a new class of off-world industries, turning space manufacturing into a commercially viable enterprise with global economic implications.

The evolution of microreactors into fully autonomous systems will be another key enabler of this vision. Powered by AI and ML algorithms, next-generation microreactors will operate independently, managing complex chemical reactions, optimizing process parameters in real time, and maintaining safety protocols without human intervention. These autonomous systems will function around the clock, ensuring a continuous supply of vital resources while also serving as environmental monitors within space habitats and manufacturing zones.

Global collaboration will be essential to the success of space-based manufacturing hubs. These platforms will act as international research centers, welcoming scientists, engineers, and innovators from across the globe to conduct cutting-edge experiments and develop advanced technologies. In doing so, they will foster a new era of cooperation, where shared knowledge and mutual ambition drive collective progress in the pursuit of sustainable space exploration.

Microreactor systems will become the cornerstone of this future essential to the infrastructure of space manufacturing, human habitation, and deep-space mission planning. Far more than compact chemical processors, they represent the embodiment of human ingenuity and determination. Their role will extend from enabling basic life support to unlocking entirely new classes of materials and applications, both in space and on Earth.

Realizing this vision will require bold investment, interdisciplinary collaboration, and sustained innovation. Governments, private enterprises, and the global scientific community must align their efforts to develop, deploy, and refine these systems. As space transitions from a frontier of exploration to a domain of creation and growth, microreactor technologies will be at the heart of our efforts to build a thriving, self-sustaining human presence beyond Earth.

We now stand at the dawn of a new era where space is no longer just a destination, but a place of innovation, production, and permanent human settlement. Microreactors will fuel this transformation, enabling us to explore deeper, build smarter, and endure longer in the vast expanse of the cosmos. They are not only tools of science and engineering but also symbols of our ambition to thrive among the stars.

CHAPTER 7

Application Examples

Microgravity, or weightlessness, offers a unique environment for material synthesis and the study of material properties. The absence of gravitational forces can lead to novel structures and properties in materials that are not achievable on Earth. Research conducted in microgravity environments, such as the International Space Station (ISS) or parabolic flight experiments, has yielded valuable insights into various material systems. The following section examines some key aspects of microgravity material synthesis and properties.

Microgravity offers a transformative environment for material synthesis, allowing scientists to explore and optimize processes that are otherwise constrained by Earth's gravity. In space, the absence of buoyancy-driven convection, sedimentation, and gravitational segregation enables uniform crystallization, precise diffusion, and controlled phase transitions. These advantages have led to breakthroughs in the production of defect-free crystals, homogeneous alloys, and uniform nanoparticles, resulting in advanced materials with superior mechanical, thermal, optical, and chemical properties. Experiments aboard platforms like the ISS have demonstrated improvements in protein crystallization for pharmaceutical research, alloy solidification for aerospace applications, and colloidal and semiconductor crystal formation for photonic and electronic devices.

Applications extend to the development of high-performance composite materials, where uniform distribution and alignment of reinforcing agents like carbon nanotubes or ceramic fibers lead to stronger, lighter, and more durable structures. In parallel, nanoparticle synthesis in

DOI: 10.1201/9781003656081-7

microgravity yields materials with enhanced surface properties, uniform size distributions, and novel functionalities relevant to drug delivery, catalysis, and quantum devices. Microgravity also allows for the production of foams and porous materials with controlled microstructures, enabling advancements in thermal insulation, filtration, energy storage, and biomedical scaffolding.

Altogether, the insights gained from microgravity experiments are revolutionizing the fields of materials science and engineering, laying the groundwork for next-generation innovations in both space exploration and Earth-based technologies. As space research continues, microgravity-enabled material synthesis will remain a cornerstone for scientific advancement and industrial application across disciplines.

FLUID DYNAMICS AND DIFFUSION

In microgravity, fluid dynamics and diffusion behavior are significantly altered. This can affect processes like crystal growth, alloy formation, and particle dispersion. Researchers can explore the fundamental principles of fluid behavior and use this knowledge to optimize material synthesis.

EXAMPLES OF FLUID DYNAMICS AND DIFFUSION APPLICATIONS IN MICROGRAVITY FOR MATERIAL SYNTHESIS

High-Quality Protein Crystal Growth

- Application: In microgravity, the absence of buoyancy-driven convection allows proteins to form larger and more uniform crystals. This is essential for structural biology, as high-quality crystals enable more precise X-ray diffraction studies.
- Example: The growth of lysozyme crystals on the ISS has shown improved clarity and size, aiding pharmaceutical research for drug design and enzyme engineering.
- Impact: These enhanced protein structures help scientists better understand diseases and design more effective treatments.

Alloy Solidification Without Gravitational Segregation

- Application: On Earth, gravity-driven convection and sedimentation often cause imperfections and inhomogeneities in alloy structures. In microgravity, these effects are minimized, allowing for more uniform mixing and solidification.

- Example: The Transparent Alloys (TRANSPARENT-1) experiment aboard the ISS studied solidification processes, yielding insights into dendritic growth patterns.
- Impact: This research enables the design of alloys with superior strength, conductivity, and thermal properties for aerospace and industrial applications.

Colloidal Suspension Stability

- Application: Colloidal suspensions (e.g., nanoparticles in liquid) are prone to sedimentation and aggregation under gravity. In microgravity, these effects are suppressed, allowing precise study of colloidal stability and self-assembly.
- Example: The Advanced Colloids Experiment (ACE) aboard the ISS has helped understand colloidal crystallization, which is relevant for photonic and pharmaceutical materials.
- Impact: Improved colloidal materials can revolutionize optical coatings, sensors, and drug delivery systems.

Controlled Diffusion in Liquid Mixtures

- Application: Microgravity eliminates gravitational convection, enabling researchers to study pure diffusion-driven mass transport in liquids. This facilitates better control of concentration gradients during material synthesis.
- Example: The Diffusion Coefficient Experiment (DCE) has provided precise measurements of diffusion rates in liquid metals and polymer solutions.
- Impact: These findings allow engineers to optimize manufacturing processes for homogeneous material structures.

Advanced Semiconductor Crystal Growth

- Application: Semiconductor crystals, such as GaAs and InP, require defect-free growth to maximize their performance in electronic and photonic devices. Microgravity reduces convection-related imperfections during crystal formation.

- Example: The CrIS (Crystal Growth of Semiconductor Materials in Microgravity) experiment demonstrated improved crystal quality in microgravity.
- Impact: Enhanced semiconductors improve performance in solar panels, LEDs, and high-speed communication systems.

Foam and Emulsion Stability

- Application: The behavior of foams and emulsions is affected by gravity, leading to drainage and phase separation. Microgravity allows scientists to study long-term stability and structure without these distortions.
- Example: The Foam Coarsening Experiment aboard the ISS provided insights into bubble dynamics and foam stability.
- Impact: These findings improve formulations for personal care products, food science, and advanced insulation materials.

Thermocapillary Convection Control

- Application: In microgravity, thermocapillary convection the movement of fluids along temperature gradients due to surface tension becomes a dominant force. This can be harnessed for precision control of material interfaces.
- Example: The Marangoni Experiment in Space (MEIS) examined thermocapillary flows in liquid bridges, revealing patterns that cannot be observed on Earth.
- Impact: Applications include advanced soldering techniques and defect-free thin film coatings.

Nanoparticle Assembly and Self-Organization

- Application: Gravity-driven sedimentation interferes with the controlled self-assembly of nanoparticles. Microgravity allows for uniform distribution and enhanced self-organization of nanoparticles.
- Example: Research on the ISS has enabled controlled nanoparticle arrangement into photonic crystals with uniform properties.
- Impact: These crystals are used in high-performance sensors, catalysts, and advanced optics.

Bioengineered Materials Formation

- Application: Microgravity enables unique interactions between biological and synthetic materials, such as scaffolding for tissue engineering or biomaterial coatings.
- Example: Microgravity studies on hydrogel formation have enabled better control over porosity and uniformity, essential for tissue engineering scaffolds.
- Impact: These advances contribute to better medical implants, drug delivery systems, and wound healing technologies.

Superconducting Materials Research

- Application: Microgravity facilitates the study of phase transitions and magnetic field effects in superconductors without gravitational disturbances.
- Example: Research in microgravity environments has demonstrated enhanced purity and defect-free formation of high-temperature superconductors.
- Impact: Improved superconductors are crucial for applications in magnetic levitation, energy storage, and advanced medical imaging systems.

Glass and Optical Fiber Formation

- Application: Gravity can induce imperfections in glass and optical fiber production. Microgravity allows precise control over the viscosity and surface tension during fiber drawing.
- Example: The Fiber Optics Manufacturing in Space (FOMS) project has successfully produced high-quality ZBLAN optical fibers with significantly fewer imperfections.
- Impact: These fibers enable better signal transmission in telecommunications and advanced medical imaging.

Microencapsulation for Drug Delivery

- Application: Microgravity facilitates the formation of uniform microcapsules, which are essential for controlled drug delivery systems.

- Example: Experiments on the ISS have produced more uniform microcapsules compared to Earth-based production.
- Impact: These advancements enable precise drug dosing and prolonged release, enhancing therapeutic efficacy.

These examples illustrate how microgravity provides a unique environment to study and optimize fluid behavior, diffusion processes, and material synthesis techniques, resulting in innovations across fields such as pharmaceuticals, aerospace, and advanced manufacturing.

CRYSTAL GROWTH

Microgravity provides an environment with reduced buoyancy-driven convection, resulting in more uniform and well-defined crystal growth. This is particularly important for semiconductor and protein crystal growth. Crystals grown in microgravity can have fewer defects and higher quality.

Applications of Crystal Growth in Microgravity

Microgravity offers a unique environment for crystal growth, minimizing the effects of buoyancy-driven convection, sedimentation, and uneven temperature gradients. This results in higher-quality crystals with fewer defects and more uniform structures. Detailed examples of how microgravity facilitates advancements in crystal growth across different fields are as follows.

Protein Crystal Growth for Drug Design

- Application: Protein crystals grown in microgravity are larger and more structurally perfect, allowing scientists to determine their molecular structure with higher precision through X-ray crystallography.
- Example: The *Lysozyme Crystal Growth Experiment (LDE)* aboard the ISS demonstrated significantly larger and more ordered protein crystals.
- Impact: Improved structural data of proteins enhance the design of targeted drugs for diseases such as cancer, Alzheimer's, and infectious diseases.

Semiconductor Crystal Growth (Gallium Arsenide (GaAs))

- Application: GaAs crystals are widely used in high-efficiency solar panels and optoelectronics. Gravity-induced imperfections can reduce their performance.
- Example: *The Advanced Semiconductor Crystal Growth Experiment* on the ISS achieved defect-free GaAs crystals.
- Impact: These high-purity crystals improve the efficiency of solar cells and laser technologies, crucial for space missions and terrestrial applications.

Zeolite Crystal Synthesis for Catalysis

- Application: Zeolites are microporous crystals used in catalysis, water purification, and gas separation. Uniformity and defect-free crystal growth are critical for their efficiency.
- Example: Zeolite crystals grown on the ISS exhibited fewer structural defects compared to those synthesized on Earth.
- Impact: Improved zeolite catalysts increase efficiency in petrochemical refining and environmental applications.

High-Quality Optical Crystals (Potassium Dihydrogen Phosphate (KDP))

- Application: KDP crystals are used in high-powered lasers for applications such as fusion energy and laser-based manufacturing.
- Example: Experiments in microgravity resulted in KDP crystals with reduced internal stress and optical defects.
- Impact: Enhanced optical crystals lead to more efficient laser systems for research and industrial applications.

Insulin Crystal Growth for Diabetes Research

- Application: Insulin crystals are essential for developing controlled-release formulations in diabetes treatment.
- Example: Insulin crystals grown in microgravity showed superior size and structural clarity, providing better insight into insulin behavior.
- Impact: Improved insulin formulations offer more stable and efficient diabetes management therapies.

Colloidal Crystal Growth for Photonic Materials

- Application: Colloidal crystals are used in photonic bandgap materials, which can manipulate light at nanoscale levels.
- Example: The *ACE* aboard the ISS demonstrated better-ordered colloidal crystal growth in microgravity.
- Impact: These materials are crucial for developing advanced optical devices, including next-generation displays and sensors.

High-Temperature Superconducting Crystals

- Application: Superconducting crystals, such as YBCO, are vital for energy-efficient power systems and magnetic levitation technologies.
- Example: Microgravity experiments allowed YBCO crystals to form with fewer impurities and better superconducting properties.
- Impact: Enhanced superconductors support applications in energy transmission, MRI systems, and maglev trains.

Glucocerebrosidase Crystals for Enzyme Replacement Therapy

- Application: Glucocerebrosidase is a critical enzyme for treating Gaucher's disease. High-quality crystals improve drug formulations.
- Example: ISS experiments produced more uniform and defect-free enzyme crystals.
- Impact: These advancements ensure more effective enzyme therapies with controlled drug delivery.

Large-Scale Silicon-Germanium (SiGe) Crystals for Microelectronics

- Application: SiGe alloys are essential in high-performance microprocessors and thermoelectric devices.
- Example: Crystal growth experiments in microgravity reduced phase separation and improved homogeneity in SiGe crystals.
- Impact: Better-quality SiGe materials improve efficiency and reliability in next-generation electronic devices.

Beta-Lactamase Crystals for Antibiotic Resistance Research

- Application: Beta-lactamase is an enzyme involved in antibiotic resistance. High-quality crystals help researchers design inhibitors.
- Example: ISS-grown beta-lactamase crystals exhibited superior clarity for X-ray diffraction analysis.
- Impact: These findings aid the development of novel antibiotics to combat drug-resistant infections.

Thin-Film Crystal Growth for Flexible Electronics

- Application: Thin crystalline films are used in flexible displays, sensors, and wearable electronics.
- Example: Research in microgravity enabled the deposition of thin films with fewer defects and better uniformity.
- Impact: These advancements support the development of durable and efficient flexible electronic devices.

Urea Crystal Formation for Medical Applications

- Application: Urea crystals play a role in medical diagnostic technologies and pharmaceutical formulations.
- Example: Experiments on the ISS demonstrated slower and more controlled urea crystal growth, resulting in uniform structures.
- Impact: Improved crystal quality enhances diagnostic accuracy and pharmaceutical stability.

Crystal growth in microgravity provides unparalleled opportunities to understand and optimize the physical and chemical processes governing crystallization. These advancements have far-reaching applications across pharmaceuticals, electronics, photonics, catalysis, and superconductivity, driving innovation and improving technologies for both space and Earth-based industries.

ALLOY FORMATION

In microgravity, alloy formation can occur with greater uniformity and control. This is advantageous for the development of advanced materials with specific mechanical, thermal, or electrical properties.

Applications of Alloy Formation in Microgravity for Advanced Material Development

In microgravity, alloy formation benefits from the absence of buoyancy-driven convection, sedimentation, and gravitational segregation. This environment allows scientists to achieve uniform composition, precise control over phase formation, and improved material properties. Twelve detailed examples illustrating how microgravity advances alloy research and application are as follows.

Directional Solidification of Alloys

- Application: Microgravity allows controlled directional solidification, minimizing buoyancy-driven convection and macro-segregation.
- Example: The *TRANSPARENT-1* experiment aboard the ISS investigated dendritic solidification in transparent alloy systems to study phase transitions.
- Impact: Improved understanding of solidification dynamics leads to alloys with tailored mechanical and thermal properties, critical for turbine blades in aerospace and power generation.

Lead-Tin (Pb-Sn) Alloy Uniformity

- Application: The Pb-Sn alloy system is widely used in soldering and electronic connections. In microgravity, segregation caused by density differences is minimized.
- Example: Experiments have shown more uniform Pb-Sn alloy structures in space.
- Impact: Enhanced mechanical strength and conductivity make these alloys more reliable in electronics and high-performance circuits.

Aluminum-Silicon (Al-Si) Casting Improvements

- Application: Al-Si alloys are valued for their lightweight and high-strength properties, commonly used in aerospace and automotive industries.
- Example: The *Solidification Experiment on Board Apollo (SEBA)* revealed reduced porosity and fewer microstructural defects in Al-Si alloys formed in microgravity.
- Impact: Improved casting techniques result in higher durability and fatigue resistance in engine components and structural parts.

Nickel-Based Superalloys

- Application: Nickel-based superalloys are used in extreme environments, such as jet engines and power turbines, where thermal stability is crucial.
- Example: Microgravity experiments have produced nickel alloys with finer and more homogeneous grain structures.
- Impact: Enhanced high-temperature performance and resistance to creep deformation expand their operational lifespan.

Iron-Nickel (Fe-Ni) Alloys for Magnetic Applications

- Application: Fe-Ni alloys are essential for magnetic shielding and transformer cores due to their soft magnetic properties.
- Example: Experiments in microgravity achieved more uniform crystal structures in Fe-Ni alloys, reducing magnetic losses.
- Impact: More efficient magnetic materials improve energy transmission and reduce losses in power grids.

Titanium-Aluminum (Ti-Al) Intermetallic Alloys

- Application: Ti-Al intermetallic compounds exhibit high-temperature strength and oxidation resistance, making them ideal for aerospace applications.
- Example: Space-based research showed reduced microstructural imperfections in Ti-Al alloys.
- Impact: Enhanced material properties improve engine components' weight-to-performance ratio and durability.

Transparent Oxide Alloys for Photonics

- Application: Transparent oxide alloys are used in lasers, optical coatings, and photonic devices.
- Example: Microgravity experiments have shown better control over phase separation, improving transparency and optical uniformity.
- Impact: These advancements contribute to high-precision optical devices and energy-efficient photonics.

Shape Memory Alloys (Nickel-Titanium (NiTi))

- Application: Shape memory alloys, such as NiTi, are used in aerospace, medical implants, and robotics for their ability to return to a pre-set shape upon heating.
- Example: Microgravity research resulted in NiTi alloys with reduced defects and improved phase transformation behavior.
- Impact: Enhanced shape memory properties improve reliability in aerospace actuators and biomedical stents.

Aluminum-Lithium (Al-Li) Lightweight Alloys

- Application: Al-Li alloys are critical for lightweight aerospace structures due to their high strength-to-weight ratio.
- Example: Experiments on the ISS demonstrated uniform lithium distribution, reducing microcracks.
- Impact: These alloys reduce fuel consumption and improve performance in aerospace applications.

Copper-Zinc (Cu-Zn) Brass Alloys for Corrosion Resistance

- Application: Cu-Zn alloys are commonly used in marine and industrial environments due to their corrosion resistance.
- Example: Space experiments achieved uniform distribution of zinc in the copper matrix.
- Impact: Improved corrosion resistance ensures longer service life in harsh environments.

Bimetallic Composites for Thermal Management

- Application: Bimetallic materials, combining two metals with different thermal expansion coefficients, are essential in aerospace and electronics.
- Example: In microgravity, the bonding interfaces of bimetallic composites showed fewer defects and better adhesion.
- Impact: These materials improve performance in thermal management systems, reducing risks of mechanical failure.

High-Entropy Alloys (HEAs)

- Application: HEAs, composed of multiple principal elements, exhibit exceptional strength, corrosion resistance, and thermal stability.
- Example: Microgravity research allowed controlled solidification and uniform distribution of elements in HEAs.
- Impact: These alloys are suitable for extreme conditions in aerospace, nuclear reactors, and high-performance tools.

Key Benefits of Alloy Formation in Microgravity

- Uniform Composition: Reduced gravitational forces prevent density-driven segregation.
- Controlled Solidification: Slower solidification allows precise control over microstructures.
- Improved Mechanical Properties: Enhanced grain structures and reduced defects lead to better strength and fatigue resistance.
- Advanced Thermal and Electrical Properties: Optimized heat conduction and electrical behavior expand industrial and technological applications.

These examples highlight how microgravity environments enable scientists to push the boundaries of materials science, leading to advanced alloys with unprecedented properties suitable for aerospace, electronics, healthcare, and sustainable technologies.

NANOPARTICLE FORMATION

Microgravity conditions can lead to the formation of nanoparticles with unique sizes, shapes, and properties. These nanoparticles can find applications in drug delivery, catalysis, and materials science.

Applications of Nanoparticle Formation in Microgravity

Microgravity conditions provide an environment with minimized sedimentation, convection currents, and buoyancy effects, allowing precise control over nanoparticle formation processes. This results in nanoparticles with unique sizes, shapes, and properties, enhancing their applications in drug delivery, catalysis, materials science, and other advanced technologies. Twelve detailed examples of how microgravity enhances nanoparticle synthesis and their applications are as follows.

Drug Delivery Nanoparticles

- Application: Nanoparticles can encapsulate drugs, allowing controlled and targeted delivery to specific tissues or organs.
- Example: In microgravity, drug-loaded nanoparticles (e.g., liposomes or polymeric nanoparticles) exhibit greater uniformity and controlled release properties.
- Impact: Improved nanoparticle formulations enhance therapeutic efficiency, reduce side effects, and allow for precise dosing in cancer and chronic disease treatments.

Catalytic Nanoparticles for Energy Conversion

- Application: Catalysts at the nanoscale improve efficiency in chemical reactions, such as hydrogen production or carbon dioxide conversion.
- Example: Platinum nanoparticles synthesized in microgravity have shown higher surface area and improved catalytic activity.
- Impact: Enhanced catalysts are essential for green energy technologies, including fuel cells and sustainable chemical synthesis.

Nanoparticles for Advanced Coatings

- Application: Nanoparticles can create coatings with anti-corrosion, anti-fouling, or self-cleaning properties.
- Example: TiO_2 nanoparticles synthesized in microgravity exhibit improved photocatalytic activity and coating uniformity.
- Impact: These coatings are used in aerospace structures, medical implants, and architectural surfaces.

Nanoparticles for Photonic Applications

- Application: Nanoparticles with specific optical properties are used in photonic crystals, sensors, and optical fibers.
- Example: Silica and gold nanoparticles produced in microgravity have exhibited consistent size and high optical quality.
- Impact: Improved optical properties enhance applications in sensors, lasers, and communication devices.

Magnetic Nanoparticles for Targeted Therapy

- Application: Magnetic nanoparticles are used in hyperthermia therapy, magnetic resonance imaging (MRI), and targeted drug delivery.
- Example: Iron oxide nanoparticles synthesized in microgravity displayed uniform magnetic properties and reduced aggregation.
- Impact: These advancements enable precise targeting and enhanced imaging for cancer therapies and diagnostics.

Nanoparticles for Water Purification

- Application: Nanoparticles such as silver or titanium dioxide can effectively remove pollutants and pathogens from water.
- Example: Silver nanoparticles produced in microgravity demonstrated higher antibacterial activity due to their uniform size and surface area.
- Impact: Enhanced water purification systems are critical for space missions and remote locations on Earth.

Quantum Dots for Display Technologies

- Application: Quantum dots are nanoscale semiconductor particles used in high-definition displays and imaging systems.
- Example: Microgravity allows precise control over quantum dot size, resulting in improved color accuracy and brightness.
- Impact: Advanced quantum dot displays provide better resolution, energy efficiency, and vibrant colors.

Nanoparticles for Aerospace Materials

- Application: Nanoparticles can reinforce materials, improving strength, heat resistance, and durability.
- Example: Aluminum oxide (Al_2O_3) nanoparticles synthesized in microgravity exhibit improved dispersion and uniformity in composite materials.
- Impact: Lightweight, high-strength materials are essential for spacecraft, satellites, and aerospace structures.

Nanoparticles for Biomedical Imaging

- Application: Nanoparticles are used as contrast agents in MRI and CT imaging for enhanced diagnostic accuracy.
- Example: Gold nanoparticles synthesized in microgravity demonstrated precise size control and increased biocompatibility.
- Impact: Improved imaging agents enhance early disease detection and non-invasive diagnostics.

Nanoparticles for Environmental Remediation

- Application: Nanoparticles can absorb or break down pollutants in soil and water.
- Example: ZnO nanoparticles formed in microgravity showed enhanced photocatalytic degradation of organic pollutants.
- Impact: These nanoparticles support eco-friendly remediation techniques for polluted environments.

Nanoparticles for Battery Technologies

- Application: Nanoparticles improve battery electrode materials, enhancing energy storage capacity and charge cycles.
- Example: Lithium cobalt oxide ($LiCoO_2$) nanoparticles synthesized in microgravity displayed uniform particle size and reduced defects.
- Impact: Enhanced battery performance is crucial for electric vehicles, portable electronics, and space applications.

Nanoparticles for Smart Textiles

- Application: Nanoparticles can impart functionalities like self-cleaning, UV protection, and antimicrobial properties to fabrics.
- Example: Silver and titanium dioxide nanoparticles produced in microgravity were integrated into textiles for improved durability and functionality.
- Impact: Smart textiles have applications in healthcare, military gear, and outdoor equipment.

Key Advantages of Nanoparticle Formation in Microgravity

- Uniform Size and Shape: Microgravity minimizes convection currents and sedimentation, enabling consistent particle formation.
- Controlled Nucleation and Growth: Better control over the crystallization process ensures precise particle properties.
- Reduced Agglomeration: Nanoparticles exhibit lower aggregation tendencies in microgravity, preserving their functional properties.
- Enhanced Surface Properties: Improved surface area and reactivity optimize performance in catalysis and material reinforcement.
- Novel Properties: Unique structural and optical properties emerge in microgravity-synthesized nanoparticles.

Microgravity environments provide unparalleled opportunities for precise nanoparticle synthesis, resulting in materials with enhanced properties and functionalities. These advancements have transformative impacts across healthcare, energy, aerospace, environmental science, and electronics, paving the way for innovative technologies both on Earth and in space exploration.

COMPOSITE MATERIALS

Microgravity enables the production of composite materials with improved dispersion and alignment of reinforcing materials (e.g., fibers or nanoparticles) within a matrix. This results in enhanced mechanical properties and structural performance.

Applications of Composite Material Production in Microgravity

Microgravity provides a unique environment for the synthesis of composite materials, minimizing gravitational effects such as sedimentation, buoyancy-driven convection, and uneven phase separation. This allows for better dispersion, alignment, and bonding of reinforcing materials (e.g., fibers, nanoparticles, or whiskers) within the matrix. These advancements result in composites with enhanced mechanical, thermal, and electrical properties, paving the way for improved structural performance in aerospace, automotive, medical, and energy sectors. Twelve detailed examples illustrating the impact of microgravity on composite material production are as follows.

Carbon Fiber-Reinforced Polymers (CFRPs)

- Application: CFRPs are widely used in aerospace and automotive industries due to their exceptional strength-to-weight ratio.
- Example: In microgravity, carbon fibers align more uniformly within the polymer matrix, reducing voids and improving fiber-matrix adhesion.
- Impact: Enhanced structural strength, durability, and fatigue resistance make CFRPs ideal for spacecraft components and lightweight aerospace structures.

Nanoparticle-Reinforced Metal Matrix Composites (MMCs)

- Application: Metal matrix composites reinforced with nanoparticles offer high strength, wear resistance, and thermal stability.
- Example: Aluminum matrix composites with uniformly dispersed SiC nanoparticles were synthesized in microgravity with fewer defects and agglomerations.
- Impact: These materials are ideal for high-performance automotive and aerospace engine components.

Ceramic-Matrix Composites (CMCs)

- Application: CMCs are essential for high-temperature applications, such as jet engines and TPSs.
- Example: In microgravity, ceramic fibers were uniformly distributed within a SiC matrix, reducing weak points and enhancing thermal shock resistance.
- Impact: Improved reliability in extreme environments makes CMCs critical for aerospace propulsion systems.

Graphene-Based Composites for Conductive Coatings

- Application: Graphene composites are used in advanced coatings for electrical conductivity, thermal management, and anti-corrosion properties.
- Example: Microgravity experiments showed improved dispersion of graphene sheets within polymer matrices, reducing agglomeration and optimizing conductive pathways.
- Impact: Enhanced coatings are ideal for spacecraft surfaces, smart textiles, and electronic devices.

Nanotube-Reinforced Polymers

- Application: Carbon nanotubes (CNTs) can significantly improve the strength, flexibility, and conductivity of polymer composites.
- Example: In microgravity, CNTs were evenly dispersed within epoxy resin, avoiding clustering and improving mechanical properties.
- Impact: These composites are used in lightweight structural components, advanced sensors, and conductive adhesives.

Biocompatible Composite Scaffolds for Tissue Engineering

- Application: Composite scaffolds are used to support cell growth and tissue regeneration in biomedical applications.
- Example: Hydroxyapatite and collagen composites synthesized in microgravity exhibited uniform distribution of bioactive materials.
- Impact: Improved scaffolds lead to better outcomes in bone regeneration and tissue repair.

Thermal Barrier Coatings (TBCs) for Aerospace Applications

- Application: TBCs protect turbine blades and aerospace structures from high temperatures and oxidation.
- Example: Ceramic-based TBCs formed in microgravity showed more consistent layer thickness and fewer microcracks.
- Impact: Enhanced thermal resistance extends the lifespan of engine components in the aerospace and energy industries.

Self-Healing Composite Materials

- Application: Self-healing composites contain microcapsules or nanofibers that release healing agents upon damage.
- Example: Microgravity studies enabled better encapsulation and distribution of healing agents within polymer matrices.
- Impact: These materials enhance durability and reduce maintenance costs in aerospace and structural applications.

Magnetic Nanocomposites for Sensors and Actuators

- Application: Magnetic nanocomposites are used in advanced sensors, actuators, and energy harvesting systems.

- Example: In microgravity, magnetic nanoparticles were evenly dispersed in polymer matrices, improving magnetic response.
- Impact: These composites are essential for precision devices, robotics, and smart materials.

Lightweight Polymer Nanocomposites for Spacecraft Structures

- Application: Lightweight composites reduce spacecraft launch mass while maintaining structural integrity.
- Example: Nanoparticle-reinforced polyethylene composites demonstrated uniform distribution and reduced defect density in microgravity.
- Impact: These materials improve spacecraft fuel efficiency and overall mission cost-effectiveness.

Transparent Nanocomposites for Optical Devices

- Application: Transparent composites are used in lenses, solar panels, and optical coatings.
- Example: ZnO nanoparticles were uniformly dispersed in polymer matrices in microgravity, preserving optical clarity.
- Impact: Improved optical performance enhances solar panel efficiency and optical sensor accuracy.

Composite Supercapacitors for Energy Storage

- Application: Nanocomposite-based supercapacitors offer high energy density and fast charge–discharge cycles.
- Example: Carbon nanofiber composites synthesized in microgravity demonstrated improved electrode uniformity and surface area.
- Impact: These advancements improve energy storage systems for space missions and portable electronics.

Key Advantages of Composite Material Formation in Microgravity

- Uniform Dispersion: Reduced gravitational forces prevent sedimentation and clustering of reinforcing materials.
- Improved Alignment: Fibers and nanoparticles can align more precisely within the matrix.

- Reduced Defects: Fewer voids, cracks, and phase separation improve structural performance.
- Optimized Interfaces: Enhanced bonding between the matrix and reinforcing materials.
- Tailored Properties: Better control over mechanical, thermal, and electrical properties.

Microgravity provides a transformative platform for synthesizing composite materials with unprecedented uniformity, alignment, and defect control. These advancements lead to stronger, lighter, and more durable materials, with applications spanning aerospace, healthcare, energy storage, and advanced manufacturing. Future space-based material research promises even more innovative composite designs with properties unattainable on Earth.

FOAMING AND POROUS MATERIALS

Microgravity facilitates the production of foams and porous materials with controlled structures and properties. These materials have applications in insulation, lightweight structures, and filtration.

Applications of Foams and Porous Materials Produced in Microgravity

Microgravity provides a unique environment for the formation of foams and porous materials by minimizing buoyancy-driven convection, gravitational drainage, and sedimentation. This enables precise control over bubble formation, pore size distribution, and material uniformity. These advancements lead to materials with enhanced thermal insulation, mechanical strength, filtration efficiency, and lightweight structural properties.

Thermal Insulation Foams for Aerospace Applications

- Application: Lightweight polymer and ceramic foams are used in spacecraft insulation and cryogenic storage systems.
- Example: Polymeric foams produced in microgravity demonstrated uniform pore size and better thermal insulation properties.
- Impact: Enhanced insulation materials reduce heat loss, minimize energy consumption, and improve the safety of space missions.

Metal Foams for Lightweight Structures

- Application: Aluminum and titanium foams are used for lightweight structural applications in aerospace and automotive industries.
- Example: In microgravity, metal foams exhibit uniform pore distribution and reduced defects during solidification.
- Impact: Improved structural strength-to-weight ratios are essential for spacecraft components and automotive frames.

Porous Filters for Water and Air Purification

- Application: Porous ceramic and polymer filters are used to remove contaminants from water and air in space stations and on Earth.
- Example: Microgravity-enabled porous filters showed more uniform pore structures, enhancing filtration efficiency.
- Impact: These advanced filters are critical for closed-loop life support systems in space and environmental remediation on Earth.

Aerogels for Energy Storage and Insulation

- Application: Aerogels are ultra-lightweight porous materials used for insulation, energy storage, and environmental cleanup.
- Example: Silica aerogels synthesized in microgravity displayed highly uniform nano-pore structures with reduced defects.
- Impact: These materials are ideal for thermal barriers in spacecraft and energy-efficient insulation systems.

Ceramic Foams for High-Temperature Applications

- Application: Ceramic foams are used in TPSs and refractory materials for high-temperature environments.
- Example: Microgravity synthesis resulted in ceramic foams with highly uniform pore structures and enhanced thermal resistance.
- Impact: These foams improve the performance and lifespan of aerospace heat shields and industrial furnaces.

Bioactive Porous Scaffolds for Tissue Engineering

- Application: Porous bioactive materials are used as scaffolds for cell growth and tissue regeneration in medical applications.

- Example: Microgravity-produced hydroxyapatite scaffolds demonstrated uniform porosity and better mechanical stability.
- Impact: Enhanced scaffolds improve outcomes in bone regeneration and wound healing therapies.

Lightweight Acoustic Insulation Foams

- Application: Acoustic foams are used in spacecraft, aircraft, and building materials to reduce noise pollution.
- Example: Polyurethane foams synthesized in microgravity exhibited highly controlled pore structures for optimized acoustic absorption.
- Impact: Improved soundproofing materials are essential for both space habitats and terrestrial buildings.

Porous Electrodes for Energy Storage Devices

- Application: Porous carbon-based electrodes are used in batteries and supercapacitors for improved charge storage capacity.
- Example: Microgravity-produced porous electrodes demonstrated uniform pore size, enhancing ion transport and charge efficiency.
- Impact: These advancements lead to higher energy densities and longer lifespans in energy storage systems.

Structural Foams for Impact Absorption

- Application: Structural foams are used in protective gear, crash-resistant vehicles, and aerospace impact shields.
- Example: Microgravity environments enabled the formation of highly uniform impact-absorbing polymer foams.
- Impact: Enhanced foams offer better energy dissipation and protection against high-impact forces.

Porous Materials for Catalysis

- Application: Porous catalysts are used in chemical reactions for fuel cells, hydrogen production, and industrial synthesis.
- Example: Microgravity-produced porous catalysts exhibited higher surface area and optimized pore distribution for catalytic activity.
- Impact: These materials improve efficiency and reduce costs in industrial catalytic processes.

Open-Cell Metal Foams for Heat Exchangers

- Application: Open-cell metal foams are used in advanced heat exchangers to enhance heat transfer efficiency.
- Example: Aluminum foams formed in microgravity showed consistent open-cell structures, optimizing fluid flow and heat exchange.
- Impact: Improved heat exchanger designs reduce energy consumption in space systems and industrial applications.

Porous Materials for Gas Storage (e.g., Hydrogen and Methane)

- Application: Porous materials like MOFs are used to store gases efficiently under high pressure.
- Example: MOFs synthesized in microgravity displayed uniform pore distribution, increasing gas adsorption capacity.
- Impact: These materials are vital for efficient hydrogen storage in fuel cells and energy systems.

Key Advantages of Producing Foams and Porous Materials in Microgravity

- Uniform Pore Size Distribution: Microgravity eliminates gravity-driven drainage, ensuring consistent pore structures.
- Minimized Defects: Controlled solidification and reduced gravitational effects prevent voids and imperfections.
- Enhanced Material Properties: Improved mechanical, thermal, and acoustic performance.
- Optimized Surface Area: Better pore structures increase reactivity and adsorption capabilities.
- Scalable Production Insights: Experiments in microgravity provide valuable data for Earth-based manufacturing techniques.

Microgravity enables the creation of foams and porous materials with unprecedented structural uniformity, reduced defects, and enhanced functionality. These advancements are revolutionizing applications in aerospace insulation, environmental filtration, biomedical scaffolds, impact absorption, and energy storage systems. Future research in space-based material synthesis will continue to push the boundaries of these technologies, offering sustainable and efficient solutions for both terrestrial and space applications.

HEAT TRANSFER AND THERMAL PROPERTIES

Understanding heat transfer mechanisms in microgravity is crucial for designing efficient thermal management systems in spacecraft and space habitats. It can also inform the development of advanced materials with tailored thermal properties.

Understanding Heat Transfer Mechanisms in Microgravity

Applications in Thermal Management and Material Development

Heat transfer in microgravity operates differently from Earth-based systems due to the absence of buoyancy-driven convection. In microgravity, heat transfer primarily relies on conduction, radiation, and controlled forced convection rather than natural convection currents. This unique environment enables scientists and engineers to study heat transfer processes with unprecedented clarity, optimize thermal management systems, and design advanced materials with tailored thermal properties.

Thermal Management in Spacecraft Electronics

- Application: Efficient heat dissipation in spacecraft electronics prevents overheating and ensures long-term reliability.
- Example: Microgravity experiments have allowed the development of heat pipes using phase-change materials (e.g., ammonia, water) optimized for low-gravity environments.
- Impact: Improved thermal control systems reduce the risk of overheating in satellites, space stations, and deep-space probes.

Phase-Change Materials (PCMs) for Heat Storage

- Application: PCMs absorb and release heat during phase transitions, making them ideal for thermal energy storage in spacecraft.
- Example: Microgravity studies have optimized PCM behavior in microcapsules, ensuring efficient heat absorption and uniform phase changes.
- Impact: Enhanced PCMs provide reliable thermal buffering in space habitats and energy systems.

Boiling Heat Transfer Mechanisms

- Application: Boiling is a crucial heat transfer mechanism for cooling systems and chemical reactors in space.

- Example: The Bubbles and Boiling Experiment (BEXUS) aboard the ISS revealed that bubble behavior in microgravity differs significantly, with bubbles growing larger and persisting longer.
- Impact: Insights enable more efficient heat exchangers and cooling systems for space-based technologies.

Advanced Heat Shield Materials for Atmospheric Re-entry

- Application: Spacecraft re-entry generates intense heat, requiring advanced heat shield materials to protect the structure.
- Example: Microgravity experiments have optimized the thermal conductivity and ablative properties of ceramic composite shields.
- Impact: Enhanced heat shields improve spacecraft safety during atmospheric re-entry.

Radiative Heat Transfer Optimization

- Application: In the absence of convection, radiative heat transfer becomes the dominant mechanism in space habitats.
- Example: Microgravity studies have refined the design of radiators and reflective coatings to maximize thermal emission efficiency.
- Impact: These improvements reduce power consumption and maintain stable temperatures in spacecraft.

Thermal Management of Life Support Systems

- Application: Space habitats require precise temperature control for air circulation, water systems, and human comfort.
- Example: Experiments on the ISS have led to improved thermal control loops using advanced fluids and nanofluids.
- Impact: Enhanced life support systems create stable, comfortable living environments in long-duration space missions.

Nanofluids for Enhanced Heat Transfer

- Application: Nanofluids (fluids containing nanoparticles) exhibit superior thermal conductivity and heat transfer properties.

- Example: Microgravity research has shown that nanoparticle distribution in fluids remains more uniform in low gravity, improving heat transfer efficiency.
- Impact: Nanofluids enhance spacecraft cooling systems and high-performance electronics.

Heat Dissipation in Microelectronics

- Application: Space-based microelectronics face significant thermal challenges due to their compact designs.
- Example: Microgravity experiments have informed the design of micro-channel heat sinks for electronic chips.
- Impact: Improved thermal management extends the lifespan and reliability of space-based electronic systems.

Insulating Materials for Space Habitats

- Application: Thermal insulation is essential to regulate temperature extremes in space habitats.
- Example: Aerogels produced in microgravity exhibit highly uniform porous structures, optimizing their insulating performance.
- Impact: Enhanced aerogels improve energy efficiency and temperature control in space habitats.

Thermal Conductivity in Composite Materials

- Application: Understanding thermal conductivity in microgravity aids the design of lightweight and thermally stable composite materials.
- Example: CFRPs with tailored thermal conductivity properties were developed in microgravity.
- Impact: These materials are ideal for spacecraft structural components exposed to thermal cycling.

Vapor–Liquid Phase Separation in Heat Pipes

- Application: Heat pipes rely on vapor–liquid phase transitions for thermal regulation in spacecraft.
- Example: Research aboard the ISS examined how vapor bubbles move and condense under microgravity, optimizing heat pipe design.

- Impact: Enhanced heat pipes improve reliability and efficiency in space thermal management systems.

Cryogenic Fluid Management for Space Propulsion

- Application: Cryogenic fuels (e.g., liquid hydrogen, liquid oxygen) require efficient heat management to prevent boil-off and fuel loss.
- Example: Microgravity studies provided insights into cryogenic fuel behavior, including heat leakage and phase transitions.
- Impact: Optimized cryogenic fuel systems improve mission success rates and reduce costs in space exploration.

Key Mechanisms Studied in Microgravity Heat Transfer Research

- Boiling and Condensation: Studying bubble formation, growth, and detachment in microgravity reveals insights into phase-change heat transfer.
- Thermal Conduction: Understanding how heat propagates through solid and composite materials without gravity-driven disturbances.
- Radiative Heat Transfer: Optimizing radiative cooling systems critical in vacuum environments.
- Vapor–Liquid Behavior: Analyzing phase separation and fluid movement in heat pipes and thermal loops.
- Nanofluid Dynamics: Studying how nanoparticles distribute in cooling fluids under microgravity.

Technological Impacts of Heat Transfer Research in Microgravity

- Energy Efficiency: Improved thermal control systems reduce energy consumption in space habitats and spacecraft.
- Enhanced Material Design: Advanced thermal properties in composites and foams optimize aerospace applications.
- Long-Duration Missions: Reliable temperature regulation supports extended missions to the Moon, Mars, and beyond.
- Reduced Thermal Stress: Better thermal shielding minimizes risks associated with rapid temperature changes in space environments.

Microgravity provides a unique environment to study heat transfer mechanisms, offering insights into boiling, conduction, radiation, and vapor–liquid dynamics. These findings have led to innovations in thermal management systems, heat-resistant materials, energy storage, and life-support technologies. Future research in microgravity promises even more advanced solutions for space exploration, satellite operations, and Earth-based thermal engineering systems.

MATERIALS FOR SPACE EXPLORATION

Research in microgravity helps in the development of materials that can withstand the extreme conditions of space, including radiation, vacuum, and temperature fluctuations. This is critical for space exploration missions and the durability of spacecraft.

Development of Space-Resilient Materials Through Microgravity Research

Microgravity provides a unique platform for studying material behavior under conditions that simulate the harsh space environment. This research enables the development of materials capable of withstanding radiation, vacuum, extreme temperature fluctuations, and mechanical stresses. These materials are essential for space exploration missions, spacecraft durability, and long-term space habitation.

Radiation-Resistant Polymers

- Application: Spacecraft and habitats require protective materials against ionizing and cosmic radiation.
- Example: Research in microgravity has allowed the synthesis of high-performance polymers with reinforced molecular structures, increasing their resistance to radiation damage.
- Impact: These materials ensure the long-lasting durability of spacecraft structures and reduce radiation exposure for astronauts.

Temperature-Resistant Composites

- Application: Spacecraft experience extreme temperature fluctuations, ranging from -150°C in shadowed areas to over 200°C in sunlight.
- Example: Carbon-carbon composites fabricated in microgravity demonstrated enhanced thermal stability and resistance to thermal cycling.
- Impact: These composites improve the integrity and lifespan of spacecraft exposed to thermal extremes.

Self-Healing Materials for Micrometeoroid Protection

- Application: Spacecraft are constantly at risk of micrometeoroid impacts, causing structural damage.
- Example: Microgravity research facilitated the development of polymer composites with self-healing properties activated by temperature or UV light.
- Impact: Self-healing materials minimize maintenance needs and enhance the durability of space structures.

Vacuum-Stable Coatings for Spacecraft Surfaces

- Application: In the vacuum of space, traditional coatings degrade rapidly due to atomic oxygen and radiation exposure.
- Example: Advanced ceramic-based coatings synthesized in microgravity displayed higher durability and lower degradation rates.
- Impact: These coatings extend the operational lifespan of spacecraft exteriors.

Transparent Radiation Shields for Observation Windows

- Application: Transparent shields are required for spacecraft observation ports, combining visibility with radiation protection.
- Example: Microgravity experiments enabled the creation of borosilicate glass composites with reduced structural defects.
- Impact: Enhanced visibility and protection ensure safe and reliable observational capabilities.

Low-Emissivity Thermal Control Films

- Application: Thermal control films regulate spacecraft temperature in vacuum and extreme solar exposure.
- Example: Microgravity research optimized thin-film deposition techniques, creating films with precise emissivity properties.
- Impact: These films improve energy efficiency and reduce temperature fluctuations in spacecraft systems.

Lightweight Alloys for Structural Integrity

- Application: Space missions require strong yet lightweight structural materials to optimize fuel efficiency and reduce launch mass.

- Example: Al-Li alloys produced in microgravity exhibited superior strength-to-weight ratios and minimal impurities.
- Impact: Lightweight alloys reduce spacecraft weight while maintaining structural integrity.

Nanostructured Radiation Shields

- Application: Nanomaterials are used to create highly effective radiation shields with lower weight and improved efficiency.
- Example: Microgravity synthesis enabled precise control over nanostructure alignment in radiation-absorbing materials like boron nitride nanotubes.
- Impact: Enhanced radiation shields protect sensitive electronics and human occupants in spacecraft.

Advanced Ablative Heat Shield Materials

- Application: Heat shields protect spacecraft during atmospheric re-entry by withstanding extreme temperatures.
- Example: Microgravity research optimized the distribution of ablative compounds in heat shield materials, improving thermal resistance.
- Impact: Safer and more efficient atmospheric re-entry systems increase mission reliability.

Flexible Solar Panel Materials

- Application: Solar panels must endure vacuum, temperature changes, and solar radiation while remaining lightweight and foldable.
- Example: Microgravity experiments enabled the development of flexible photovoltaic materials with enhanced efficiency and reduced brittleness.
- Impact: These materials increase energy generation capacity in space missions.

Atomic Oxygen-Resistant Materials

- Application: In LEO, atomic oxygen corrodes spacecraft surfaces, degrading structural and functional materials.
- Example: Microgravity research has led to the synthesis of ceramic-metal composites resistant to atomic oxygen attack.

- Impact: Improved resistance ensures longer operational lifetimes for spacecraft in LEO.

Shape Memory Alloys for Adaptive Structures

- Application: SMAs (e.g., Nickel-Titanium [NiTi]) allow spacecraft components to adapt to environmental changes or repair damage.
- Example: Microgravity fabrication of NiTi alloys improved their phase transformation behavior and durability.
- Impact: These alloys enable adaptive structures, such as deployable antennas and self-repairing mechanisms in satellites.

Key Advantages of Space-Resilient Materials Developed in Microgravity

- Radiation Resistance: Enhanced shielding properties protect both spacecraft and astronauts from harmful cosmic rays.
- Thermal Stability: Materials maintain structural integrity across wide temperature fluctuations.
- Self-Healing Capabilities: Reduced maintenance requirements and extended material lifespans.
- Atomic Oxygen Resistance: Prolonged durability in LEO environments.
- Lightweight Design: Optimization of weight-to-strength ratios reduces mission costs and fuel usage.
- Vacuum Stability: Materials maintain performance in near-perfect vacuum conditions.
- Improved Surface Coatings: Enhanced protective coatings reduce corrosion and surface degradation.

Technological Impacts on Space Exploration

- Longer Mission Durations: Materials can endure long-term exposure to harsh space environments.
- Improved Spacecraft Efficiency: Reduced mass and enhanced performance optimize fuel efficiency.
- Enhanced Safety for Astronauts: Radiation and thermal protection ensure safer habitats.

- Advanced Space Structures: Deployable and self-repairing materials enable innovative spacecraft designs.
- Sustainable Space Habitats: Materials support closed-loop life support systems and long-term orbital habitats.

Microgravity research plays a pivotal role in advancing the development of space-resilient materials. These materials are essential for addressing the thermal, mechanical, and radiation challenges of space exploration. Innovations in polymers, composites, nanostructures, and adaptive materials contribute to the durability, efficiency, and safety of spacecraft, habitats, and space-based infrastructure. As humanity pushes toward deep-space missions and planetary colonization, these advanced materials will remain the foundation for sustainable and reliable exploration.

BIOMATERIALS AND TISSUE ENGINEERING

In microgravity, researchers can explore the behavior of biological materials, such as cells and tissues, which can have applications in regenerative medicine and tissue engineering.

3D Tissue Growth Without Scaffolding

- Example: In microgravity, cells can grow and assemble into complex three-dimensional (3D) tissue structures without relying on scaffolds for support. For instance, the ISS Tissue Chips Project demonstrated the growth of liver and cardiac tissues in space, where cells self-assembled into organized structures more effectively than on Earth.
- Impact: These self-assembled 3D tissue models are more physiologically accurate, enabling better simulations of human organs. This leads to improved drug testing, disease modeling, and potential advancements in organ transplantation, reducing the dependency on donor organs.

Stem Cell Differentiation Control

- Example: Research on the Microgravity Expanded Stem Cell Project revealed that stem cells in space exhibit altered differentiation pathways. For example, mesenchymal stem cells (MSCs) in microgravity show the enhanced ability to become bone and cartilage tissue cells.

- Impact: Understanding and controlling these differentiation pathways can lead to improved stem cell therapies for diseases such as osteoporosis, cartilage damage, and spinal cord injuries. It also opens the door to the efficient production of specific cell types for medical treatments.

Accelerated Aging Models

- Example: Astronauts on long-duration space missions experience rapid bone loss and muscle atrophy, mimicking age-related diseases such as osteoporosis and sarcopenia. Researchers on the ISS use cell cultures and tissue samples to study these accelerated aging effects.

- Impact: By studying how cells and tissues age more rapidly in microgravity, scientists can identify key molecular targets for treatments. This accelerates the development of therapies for age-related conditions and helps improve life quality for aging populations on Earth.

Drug Testing on Tissue Models

- Example: In the Organs-on-Chips Space Project, tissue chips containing heart, liver, and kidney cells are exposed to microgravity to better understand how they respond to medications. The tissues behave more like they do in the human body due to their 3D organization.

- Impact: These experiments improve the accuracy of drug testing, reducing reliance on animal testing and expediting drug discovery processes. It also ensures drugs are safer and more effective before they reach human trials.

Biomaterial Fabrication

- Example: In microgravity, bio-inks used for 3D bioprinting can form more precise and uniform tissue scaffolds. NASA's bioprinting projects have successfully created cartilage and vascular tissue structures in space.

- Impact: Biomaterials created in space demonstrate enhanced properties, such as improved mechanical strength and biological compatibility. These materials are essential for constructing functional tissue scaffolds and bioengineered organs.

Understanding Bone Loss Mechanisms

- Example: In microgravity, astronauts lose bone density at a rate of 1–2% per month. Researchers on the ISS use osteoblast and osteoclast cell cultures to study how microgravity disrupts bone regeneration and resorption balance.
- Impact: Insights gained from these studies help develop treatments for osteoporosis and other bone degenerative diseases. They also aid in creating pharmaceutical interventions to prevent bone loss during long-term space missions.

Wound Healing Research

- Example: In microgravity, researchers observed changes in fibroblast migration and collagen deposition, two critical processes for wound healing. Studies on the ISS have shown that fibroblasts respond differently to growth factors in space.
- Impact: This research helps optimize treatments for chronic wounds, burns, and diabetic ulcers. It also supports the development of advanced wound dressings and growth factor therapies for improved healing on Earth.

Vascular Network Formation

- Example: Researchers have found that endothelial cells, which form blood vessels, can grow into more organized vascular networks in microgravity. These networks are critical for delivering nutrients and oxygen to engineered tissues.
- Impact: Enhanced vascularization supports the growth of larger and more complex bioengineered tissues, bringing us closer to the reality of functional bioartificial organs suitable for transplantation.

Immune System Response Studies

- Example: In microgravity, immune cells such as T-cells and macrophages exhibit altered activity and signaling patterns. The NASA Immuno Project studied how these changes impact immune function in space.
- Impact: Understanding these alterations can help develop treatments for immune-related disorders and improve vaccines. It also ensures better health protocols for astronauts during long-duration space missions.

Organ-on-a-Chip Technologies

- Example: Devices like Heart-on-a-Chip and Kidney-on-a-Chip have been deployed on the ISS to mimic the behavior of human organs in microgravity. These chips allow precise monitoring of tissue responses to drugs and environmental changes.
- Impact: Organ-on-a-chip technologies in space enable highly accurate modeling of diseases and drug responses, reducing dependence on traditional animal models and expediting therapeutic development.

Protein Crystallization for Regenerative Medicine

- Example: In microgravity, proteins can form larger and more uniform crystals, allowing researchers to better analyze their structures. For example, Lysosomal Enzyme Crystallization experiments have optimized enzyme therapies for rare diseases.
- Impact: Improved protein structures enable the design of highly specific drugs and therapies, including those critical for tissue repair and regeneration.

Space-Grown Stem Cell Therapies

- Example: Stem cells grown in microgravity environments show unique properties, such as enhanced proliferation and stability. For example, stem cells harvested and expanded in space have been studied for treating radiation-induced tissue damage.
- Impact: Space-grown stem cells could serve as a valuable resource for cell-based therapies on Earth, offering more robust and effective treatments for organ damage and degenerative diseases.

Cartilage Tissue Engineering

- Example: Researchers have successfully grown cartilage tissue in microgravity using 3D bioprinting. In space, cartilage cells can organize into uniform, functional cartilage structures without gravitational stress flattening them.
- Impact: This research advances treatments for cartilage damage caused by injuries or arthritis. The results could enable better therapies, including bioprinted cartilage implants, to restore joint function.

Muscle Atrophy Prevention and Regeneration

- Example: Astronauts experience rapid muscle atrophy in space due to microgravity. Studies on muscle cell cultures have shown altered gene expression affecting muscle growth and repair.
- Impact: Understanding these mechanisms helps design targeted therapies for muscle-wasting diseases like muscular dystrophy and sarcopenia, as well as preventive measures for astronauts on long-duration space missions.

Tumor Growth and Cancer Cell Behavior

- Example: In microgravity, cancer cells grow in 3D tumor spheroids, mimicking their growth in the human body. NASA's Microgravity Oncology Program studies these tumor models to better understand cancer progression.
- Impact: Insights from microgravity research could lead to novel cancer treatments and targeted drug delivery mechanisms.

Improved Artificial Blood Vessel Design

- Example: Microgravity has enabled the growth of endothelial cells into more organized, tube-like structures that closely resemble blood vessels. Experiments have successfully demonstrated the self-assembly of vascular tissues.
- Impact: These developments are critical for creating artificial blood vessels, improving outcomes in surgeries requiring vascular grafts, and supporting engineered tissue implants.

Biofilm Growth for Wound Healing

- Example: Research has shown that biofilms groups of bacteria and other microorganisms form differently in microgravity. Scientists are studying biofilms on biocompatible surfaces to understand wound healing and infection resistance.
- Impact: This knowledge can optimize antimicrobial wound dressings and promote faster recovery from chronic wounds and burns.

Microgravity-Enhanced Bone Tissue Engineering

- Example: In space, osteoblasts (bone-forming cells) and osteoclasts (bone-resorbing cells) behave differently. Researchers have studied bone scaffold implants created with hydroxyapatite and collagen to mimic natural bone structures.

- Impact: These insights are paving the way for better bone graft materials, which could revolutionize treatments for fractures, osteoporosis, and skeletal injuries.

Optimizing Biomimetic Scaffolds

- Example: In microgravity, biomimetic scaffolds engineered structures mimicking natural tissue architecture can be formed with greater precision and porosity. Researchers have developed improved polymer scaffolds for bone and cartilage regeneration.
- Impact: These advanced scaffolds support cell growth, nutrient transport, and tissue regeneration, accelerating progress in bioengineered organs and implants.

Studying Immune Cell Suppression Mechanisms

- Example: Studies on the ISS Immune Project revealed that T-cells and B-cells behave differently in space, showing reduced activation and proliferation.
- Impact: Insights into immune cell behavior in space help develop therapies for immune suppression disorders, autoimmune diseases, and organ transplantation immunosuppression management.

Modeling Neurodegenerative Diseases

- Example: Brain organoids (miniature brain-like structures grown from stem cells) have been cultured in microgravity. These organoids develop more naturally in space, avoiding distortions caused by Earth's gravity.
- Impact: Better brain models enable researchers to study neurodegenerative diseases such as Alzheimer's and Parkinson's, potentially leading to innovative therapies.

Studying Tissue Responses to Radiation

- Example: In space, biological tissues are exposed to higher radiation levels. Researchers have studied the effects of radiation on cellular DNA repair and tissue regeneration processes.

- Impact: Understanding radiation effects on human tissues aids in designing countermeasures for astronauts and informs cancer radiation therapy strategies on Earth.

Research in regenerative medicine and tissue engineering in microgravity is primarily conducted aboard dedicated space platforms like the ISS. The ISS serves as a state-of-the-art orbital laboratory equipped with specialized equipment such as bioreactors, 3D bioprinters, tissue chips, and organ-on-a-chip devices. Collaborative programs between space agencies (NASA, ESA, Roscosmos, JAXA) and private spaceflight companies (e.g., SpaceX, Blue Origin, and Axiom Space) have significantly expanded access to microgravity environments for researchers worldwide.

Ground-based analog systems, such as clinostats and rotating wall bioreactors, are also used to simulate microgravity conditions for preliminary testing before spaceflight experiments. Furthermore, parabolic flights and suborbital missions offer short-duration microgravity exposure for initial trials. These diverse work environments allow scientists to study cell behavior, tissue formation, drug testing, and biomaterial performance under altered gravitational forces.

In addition to the ISS, emerging platforms such as commercial space stations (e.g., Orbital Reef) and proposed Lunar and Martian habitats will play a critical role in advancing tissue engineering research. As humanity moves toward long-duration space missions, understanding how biological systems behave in these unique conditions will become increasingly critical for both space health and terrestrial applications.

Future Directions in Microgravity Bioengineering Research

The future of regenerative medicine and tissue engineering in microgravity holds immense promise. Researchers are poised to address several critical challenges and opportunities in the coming decades.

Biofabrication of Complex Organs

While significant progress has been made in growing tissues, constructing fully functional organs with integrated vascular systems remains a challenge. Future research aims to leverage microgravity for bioprinting entire organs, such as hearts, kidneys, and livers, for transplantation.

Stem Cell Therapy Optimization

Microgravity's effect on stem cell proliferation and differentiation pathways will be further explored to optimize therapies for degenerative diseases, organ damage, and aging-related disorders.

Radiation and Tissue Response

With prolonged space travel becoming more common, understanding how radiation impacts tissue health and regeneration will be critical. Research will focus on developing radiation-resistant biomaterials and therapies to protect astronauts and patients on Earth.

Artificial Blood Vessels and Tissue Vascularization

Building fully vascularized tissues remains a hurdle for tissue engineering. Microgravity provides an ideal environment for studying endothelial cell growth and creating functional blood vessel networks within engineered tissues.

Commercialization of Space Biotechnology

As space becomes more accessible to private industries, we will likely see commercial production of bioengineered tissues, drugs, and biomaterials in microgravity environments, accelerating therapeutic advancements.

Personalized Medicine in Space

Microgravity experiments will enable personalized medicine approaches, where patient-specific tissues and organoids are grown in space and returned to Earth for tailored treatments.

The intersection of microgravity, biotechnology, and regenerative medicine is revolutionizing our ability to understand and manipulate biological systems. Current research platforms, such as the ISS, provide invaluable opportunities for groundbreaking discoveries, while upcoming commercial space stations and deep-space missions will expand these possibilities. The insights gained from microgravity research are not limited to space applications they are transforming healthcare, drug development, and tissue engineering here on Earth. As we continue to explore the final frontier, the knowledge derived from these experiments will undoubtedly pave the way for healthier lives, extended lifespans, and unprecedented medical breakthroughs both on and beyond our planet.

CHAPTER 8

Fundamental Science

FUNDAMENTAL SCIENCE

Microgravity environments allow scientists to conduct fundamental studies on material behavior, phase transitions, and physical phenomena that are otherwise masked by gravity on Earth.

Fundamental Studies of Material Behavior, Phase Transitions, and Physical Phenomena in Microgravity

Microgravity offers a unique environment for conducting fundamental research on materials and physical phenomena, free from the masking effects of gravitational forces such as buoyancy, sedimentation, and convection. These studies provide deeper insights into phase transitions, crystal growth, fluid dynamics, thermal transport, and material stability, leading to breakthroughs in material science and engineering.

Understanding Phase Transitions in Alloys

- Application: Phase transitions (e.g., solidification, melting) in metal alloys are critical for producing materials with specific mechanical properties.
- Example: The TRANSPARENT-1 experiment on the ISS allowed scientists to observe dendritic solidification patterns in alloys in real time without the effects of buoyancy-driven convection.
- Impact: Improved understanding of phase transitions helps optimize casting techniques, resulting in stronger and more durable alloys for aerospace and industrial applications.

DOI: 10.1201/9781003656081-8

Crystal Nucleation and Growth in Pure Materials

- Application: Crystal nucleation and growth determine the structural integrity and functional properties of materials, such as semiconductors and pharmaceuticals.
- Example: Protein and semiconductor crystals grown in microgravity demonstrated slower growth rates, fewer defects, and improved structural clarity.
- Impact: Higher-quality crystals improve semiconductor device performance and drug design.

Diffusion-Driven Mass Transport

- Application: Diffusion controls how atoms, molecules, or particles move through materials and fluids, affecting material properties and reaction rates.
- Example: The DCE studied diffusion in liquid metals in microgravity, isolating diffusion from convection effects.
- Impact: These insights optimize metallurgical processes, battery design, and advanced fluid dynamics models.

Thermocapillary Flow in Liquids

- Application: Thermocapillary convection (fluid movement caused by temperature-induced surface tension gradients) plays a significant role in thin-film coating and crystal growth.
- Example: The MEIS investigated thermocapillary flows in liquid bridges.
- Impact: Understanding thermocapillary dynamics improves coating technologies, defect-free film deposition, and microfluidic device design.

Vapor–Liquid Phase Separation

- Application: Vapor–liquid transitions are fundamental in heat pipes, refrigeration systems, and cryogenic storage technologies.
- Example: The TPF aboard the ISS examined bubble formation and condensation in microgravity.
- Impact: Improved phase separation processes enhance thermal management systems in satellites and spacecraft.

Colloidal Suspension Behavior

- Application: Colloidal suspensions are used in advanced materials, food science, and pharmaceuticals.
- Example: The ACE studied particle self-organization and crystallization in microgravity.
- Impact: Insights enable the design of photonic materials, stable emulsions, and drug delivery systems.

Thermal Conductivity of Nanomaterials

- Application: Understanding thermal transport in nanomaterials is essential for energy devices and heat management systems.
- Example: Microgravity studies have shown how nanofiber alignment impacts thermal conductivity in polymer composites.
- Impact: Optimized nanocomposites improve heat dissipation in microelectronics and spacecraft thermal shielding.

Foam Stability and Structure

- Application: Foams are used in thermal insulation, impact absorption, and lightweight structural materials.
- Example: The Foam Coarsening Experiment analyzed how foam structure evolves over time without gravitational drainage.
- Impact: Improved foam materials enhance aerospace insulation, packaging, and biomedical implants.

Solid–Liquid Interface Dynamics

- Application: Interface dynamics during solidification influence the formation of microstructures in cast materials.
- Example: In microgravity, researchers observed stable solid–liquid interfaces without convection-related disturbances.
- Impact: Enhanced control over microstructures improves mechanical properties in metallic and polymeric materials.

Magnetic and Superconducting Properties

- Application: Studying magnetic and superconducting materials in microgravity helps isolate intrinsic properties from gravitational disturbances.

- Example: Research on high-temperature superconductors in microgravity revealed new insights into phase transitions and magnetic flux pinning.
- Impact: Improved superconducting materials enhance magnetic levitation, power transmission, and medical imaging systems.

Glass Transition Studies

- Application: Understanding the glass transition process is critical for developing amorphous materials with tailored mechanical and optical properties.
- Example: In microgravity, glass-forming liquids were observed to undergo transitions without gravitational stresses.
- Impact: Advanced amorphous materials improve optical lenses, display technologies, and medical implants.

Combustion and Flame Dynamics

- Application: Understanding combustion in microgravity helps optimize fuel efficiency and fire safety in confined spaces.
- Example: The MCE revealed unique flame shapes and burn rates in low-gravity environments.
- Impact: Improved combustion models support cleaner energy technologies and fire safety measures in spacecraft.

Key Advantages of Fundamental Studies in Microgravity

- Isolation of Physical Phenomena: Microgravity removes gravitational masking effects, revealing fundamental processes.
- Controlled Environment: Precise control over experimental conditions allows reproducible results.
- Minimized Convection: Reduced fluid motion enables better observation of diffusion and nucleation processes.
- Enhanced Resolution: Researchers can observe material and fluid behavior at microscopic and atomic levels.
- Validation of Theoretical Models: Experimental data in microgravity refine predictive material science models.

Applications and Broader Impacts

- Aerospace: Improved lightweight alloys, thermal management systems, and protective coatings for spacecraft.
- Pharmaceuticals: High-quality protein crystals for drug design and precision medicine.
- Electronics: Optimized semiconductors and thermally conductive nanomaterials for advanced devices.
- Energy Systems: Enhanced phase-change materials, superconductors, and catalytic processes.
- Environmental Technologies: Improved foams, filters, and nanocomposites for water and air purification.

Microgravity provides an unparalleled platform for studying fundamental material behavior, phase transitions, and physical phenomena without gravitational interference. These studies offer insights into crystal growth, thermal transport, fluid dynamics, and phase separations, leading to innovations in aerospace engineering, pharmaceuticals, nanotechnology, and sustainable energy systems. As humanity advances into long-duration space missions and planetary colonization, these findings will remain critical for developing technologies that can thrive in extreme and resource-limited environments.

COMMERCIAL AND INDUSTRIAL APPLICATIONS

Insights gained from microgravity research can have practical applications in various industries, including aerospace, automotive, electronics, and healthcare.

Practical Applications for Microgravity Research Across Industries

Research conducted in microgravity environments provides valuable insights into materials, fluids, biological processes, and thermal management systems. These findings translate into practical applications across multiple industries, enhancing product performance, reliability, and efficiency.

Aerospace Industry: Advanced Lightweight Alloys

- Application: Microgravity enables the development of high-strength, lightweight alloys with improved microstructural uniformity and reduced defects.

- Example: Al-Li alloys synthesized in microgravity exhibit superior strength-to-weight ratios and fatigue resistance.
- Impact: Enhanced lightweight alloys reduce fuel consumption, improve spacecraft efficiency, and enable the construction of more durable aerospace components.

Automotive Industry: Heat-Resistant Composite Materials

- Application: Microgravity research on heat-resistant composites leads to better thermal management in engine components.
- Example: CMCs developed in microgravity display exceptional thermal stability and reduced thermal expansion.
- Impact: These materials improve fuel efficiency and reliability in high-performance automotive engines.

Electronics Industry: High-Purity Semiconductor Crystals

- Application: Defect-free semiconductor crystals are critical for high-speed electronics and energy-efficient devices.
- Example: GaAs and InP crystals grown in microgravity have fewer imperfections and better optical properties.
- Impact: These semiconductors are used in solar panels, lasers, and next-generation microelectronics.

Healthcare Industry: Protein Crystal Growth for Drug Design

- Application: Protein crystals grown in microgravity are larger and more uniform, enabling precise drug structure analysis.
- Example: High-quality insulin crystals produced in space have led to improved formulations for diabetes treatment.
- Impact: Enhanced pharmaceutical formulations improve drug efficacy, reduce side effects, and enable targeted therapies.

Energy Sector: Nanomaterials for Battery Technologies

- Application: Nanomaterials synthesized in microgravity demonstrate improved uniformity, conductivity, and energy storage capacity.

- Example: Lithium-ion batteries incorporating microgravity-developed nanomaterials exhibit better charge retention and longer lifespans.
- Impact: Advanced batteries are crucial for renewable energy storage, electric vehicles, and portable devices.

Construction Industry: Aerogels for Insulation

- Application: Microgravity facilitates the production of ultra-lightweight aerogels with highly uniform pore structures for superior insulation.
- Example: Silica aerogels developed in microgravity demonstrate exceptional thermal resistance and low thermal conductivity.
- Impact: These materials are ideal for energy-efficient building insulation and spacecraft thermal shielding.

Biotechnology: Tissue Engineering Scaffolds

- Application: Microgravity research supports the development of biocompatible porous scaffolds for tissue engineering.
- Example: Hydroxyapatite scaffolds produced in space exhibit uniform porosity, promoting better cell growth and tissue regeneration.
- Impact: These scaffolds enhance medical implants, wound healing, and tissue repair therapies.

Telecommunications: Optical Fiber Manufacturing

- Application: Microgravity enables the production of optical fibers with minimal defects and higher light transmission efficiency.
- Example: ZBLAN optical fibers manufactured in microgravity display superior signal clarity and reduced signal loss.
- Impact: These fibers improve the performance of telecommunications networks and medical imaging systems.

Environmental Science: Nanoparticle-Based Filtration Systems

- Application: Microgravity studies on nanoparticle dispersions improve water and air filtration systems.

- Example: Silver nanoparticles synthesized in microgravity exhibit enhanced antimicrobial properties and filtration efficiency.
- Impact: Advanced filtration systems are used in water treatment plants, air purification systems, and medical sterilization devices.

Food and Beverage Industry: Stable Emulsions and Foams

- Application: Understanding fluid dynamics and colloidal stability in microgravity helps create better emulsions and foams.
- Example: Food emulsions studied in microgravity exhibit improved stability and uniform texture.
- Impact: These insights optimize the production of dairy products, sauces, and carbonated beverages.

Robotics and Smart Materials: Shape Memory Alloys (SMAs)

- Application: Microgravity research has refined the properties of SMAs for robotics and adaptive structures.
- Example: NiTi alloys synthesized in space demonstrate precise shape recovery and durability.
- Impact: SMAs are used in aerospace actuators, robotic arms, and biomedical implants.

Space Tourism: Enhanced Radiation Shielding

- Application: Radiation-resistant materials developed in microgravity protect passengers and crew during space travel.
- Example: Nanostructured radiation shields created in space offer better absorption and lower weight.
- Impact: These shields improve safety and comfort in commercial space tourism ventures.

CHAPTER 9

Key Advantages of Microgravity-Enabled Applications Across Industries

- Enhanced Material Properties: Reduced defects, improved homogeneity, and superior structural integrity.
- Precision in Manufacturing: Uniform crystal growth, better particle dispersion, and consistent phase transitions.
- Increased Efficiency: Optimized heat transfer, energy storage, and fluid behavior.
- Innovative Solutions: Novel material properties and processes applicable across diverse sectors.
- Long-Term Sustainability: Environmentally friendly solutions for water purification, insulation, and renewable energy.

Microgravity offers an extraordinary environment for scientific exploration and technological advancement by minimizing the influence of gravity-driven forces such as sedimentation, convection, and buoyancy. In this weightless setting, materials can be processed in ways that are impossible

DOI: 10.1201/9781003656081-9

on Earth, leading to enhanced properties like reduced defects, improved homogeneity, and superior structural integrity. On Earth, gravity often introduces impurities and structural inconsistencies during material solidification or crystallization processes. In microgravity, however, particles and molecules remain evenly distributed, preventing density-driven separation and allowing materials to form with a higher degree of uniformity. For example, aerospace engineers can create lightweight alloys with unmatched strength and durability, while biomedical scientists can produce advanced biomimetic materials for implants that integrate seamlessly with human tissues. Similarly, the electronics industry benefits from ultra-pure semiconductors, which are crucial for improving the performance and efficiency of electronic devices. These enhanced material properties represent a fundamental advantage of microgravity-based manufacturing, laying the foundation for breakthroughs in multiple sectors.

Precision in manufacturing is another area where microgravity excels, enabling processes that yield higher-quality results compared to those on Earth. Crystal growth, a critical process in both pharmaceuticals and semiconductor manufacturing, becomes significantly more uniform in microgravity due to the absence of convection currents and gravitational inconsistencies. This uniformity results in crystals with fewer defects and greater structural purity, which is particularly important for pharmaceutical drug design and production. Likewise, optical fibers manufactured in space exhibit near-perfect internal structures, reducing signal loss and significantly improving performance in telecommunications systems. Particle dispersion in composite materials is also optimized in microgravity, ensuring consistent distribution without gravitational settling. This level of precision in material growth and particle alignment is difficult, if not impossible, to achieve under terrestrial conditions, positioning microgravity as an indispensable environment for advanced manufacturing.

Efficiency is significantly enhanced in microgravity across processes like heat transfer, fluid dynamics, and energy storage. On Earth, gravity-driven convection causes uneven heat distribution in thermal systems, creating inefficiencies in both industrial and technological applications. In microgravity, heat transfer occurs more uniformly, improving the performance of thermal management systems in both space and terrestrial applications. Similarly, fluid behavior can be more precisely controlled without the interference of buoyancy forces, allowing for innovations in microfluidics and chemical processing. In energy storage technologies, such as batteries and supercapacitors, microgravity enables uniform deposition of active materials, leading to better performance and longer operational lifespans. These advancements are especially valuable in applications requiring reliability and precision, including energy storage systems for spacecraft, renewable energy grids, and portable electronic devices.

One of the most exciting outcomes of microgravity research is its ability to enable entirely innovative solutions that transcend traditional industrial limitations. Microgravity allows scientists and engineers to explore novel material properties and manufacturing techniques that cannot be replicated on Earth. For instance, space-grown protein crystals are revolutionizing pharmaceutical research by enabling the development of more effective treatments for diseases such as cancer and Alzheimer's. Self-assembling nanomaterials, another promising area of research, leverage the weightless environment to form intricate and highly functional structures without gravitational interference. These innovations often have broad, cross-sector applicability, from biomedical engineering and telecommunications to advanced manufacturing and environmental technologies. The discoveries made in microgravity continue to generate ripple effects, driving innovation in sectors that extend far beyond their initial experimental scope.

Microgravity also plays a significant role in advancing long-term sustainability by enabling environmentally friendly technologies and resource-efficient systems. Water purification systems designed for space missions are being adapted for terrestrial use, providing clean drinking water to remote or disaster-stricken areas. Similarly, advanced insulation materials developed in microgravity have found applications in energy-efficient construction, reducing heating and cooling demands in buildings and industrial facilities. In the renewable energy sector, microgravity research has led to the development of highly efficient solar panels and next-generation energy storage technologies. These space-tested systems not only reduce resource consumption but also minimize environmental impact, aligning with global efforts to address climate change and promote sustainable development. As industries increasingly adopt microgravity-enabled technologies, the potential for creating greener, more sustainable processes grows exponentially.

Industries ranging from aerospace and pharmaceuticals to telecommunications and environmental engineering are poised to benefit from microgravity-enabled research and manufacturing. In aerospace, lightweight and durable alloys developed in space improve the performance and efficiency of spacecraft and aviation technologies. Pharmaceuticals gain from space-grown protein crystals, which provide higher-resolution structures for drug development and reduce production inefficiencies. In telecommunications, space-manufactured optical fibers deliver superior signal clarity, enabling faster and more reliable global communication networks. Environmental engineering benefits from water recycling systems and insulation technologies initially designed for space habitats. These diverse applications highlight the cross-industry versatility of microgravity technologies and underscore their potential to drive meaningful advancements in science, technology, and society.

Beyond immediate industrial benefits, microgravity research fosters a culture of innovation and scientific discovery that drives global progress. Scientists working in space-based laboratories on platforms like the ISS are uncovering insights that reshape our understanding of physics, chemistry, and biology. These insights not only contribute to scientific knowledge but also fuel practical applications with real-world benefits. For example, research into fluid dynamics in microgravity has improved fuel systems in spacecraft while also informing designs for more efficient industrial mixing processes on Earth. Likewise, thermal control technologies developed for space habitats are now being used to enhance the efficiency of cooling systems in terrestrial manufacturing plants. The knowledge gained from microgravity experiments represents a bridge between fundamental science and applied technology, accelerating innovation across disciplines.

Looking ahead, continued advancements in microgravity research are expected to have far-reaching implications for humanity. As private space companies and government agencies expand their presence in space, the accessibility of microgravity platforms will increase, opening new opportunities for research and development. Emerging fields such as space-based biomanufacturing, advanced composites, and nanotechnology are poised to revolutionize multiple industries. Furthermore, the commercialization of microgravity research will drive economic growth by creating new markets for space-enabled products and technologies. This expansion represents a transformative phase in industrial history, where space becomes not just a realm of exploration but also a cornerstone for sustainable innovation and economic development.

The advantages of microgravity-enabled applications span enhanced material properties, precision in manufacturing, increased efficiency, innovative solutions, and long-term sustainability. These benefits drive technological progress, addressing long-standing challenges in sectors ranging from healthcare and aerospace to renewable energy and telecommunications. As space research becomes more accessible through international collaborations and private sector involvement, microgravity will continue to serve as a vital platform for pioneering advancements. The technologies developed in microgravity not only improve life on Earth but also lay the groundwork for humanity's long-term presence in space, highlighting the profound and far-reaching impact of this unique scientific environment.

CROSS-INDUSTRY TECHNOLOGICAL IMPACTS

- Aerospace: Lightweight, durable materials and advanced heat shields.
- Automotive: Improved engine efficiency and fuel economy.
- Electronics: High-performance semiconductors and optical fibers.
- Healthcare: Advanced pharmaceuticals and tissue engineering scaffolds.
- Energy: Better energy storage systems and insulation materials.
- Environmental Science: Cleaner water and air through efficient filtration.

Microgravity has emerged as a powerful platform for innovation across multiple industries, offering an environment free from the gravitational forces that dominate physical and chemical processes on Earth. By removing constraints such as sedimentation, convection, and buoyancy, microgravity enables the creation of advanced materials, precise manufacturing processes, and groundbreaking technologies. In the aerospace industry, microgravity research has led to the development of lightweight yet durable materials capable of withstanding extreme conditions. These materials, including high-strength alloys and composites, contribute to reducing the weight of spacecraft and improving fuel efficiency, allowing missions to carry more payload or extend their operational range. Advanced heat shield technologies, developed microgravity experimentation, offer superior thermal resistance and better protection during atmospheric re-entry, ensuring mission safety and reducing maintenance requirements for reusable spacecraft.

The automotive sector also benefits significantly from microgravity-enabled research, particularly in the development of improved engine efficiency and fuel economy. Microgravity facilitates precise control over combustion processes and the behavior of fuel droplets, allowing engineers to optimize combustion chamber designs for reduced emissions and increased efficiency. Additionally, lightweight materials developed in space environments can be adapted for use in vehicle frames and engines, reducing overall weight and enhancing fuel economy without compromising structural integrity. Innovations in thermal management systems, inspired by space-based heat transfer research, also contribute to more efficient cooling systems in electric and hybrid vehicles, improving battery lifespan and performance.

In electronics, microgravity has unlocked the potential for creating ultrapure semiconductors and high-performance optical fibers. On Earth, gravity-driven convection and impurities often disrupt the uniform growth of semiconductor crystals, resulting in defects that degrade electronic performance. In microgravity, however, crystal growth occurs with unparalleled uniformity, allowing the production of semiconductors with near-perfect structures. These materials are essential for manufacturing faster, more efficient microprocessors and advanced electronic devices. Similarly, optical fibers produced in microgravity environments exhibit fewer internal flaws and improved clarity, allowing for faster data transmission over longer distances with minimal signal loss. These advancements have far-reaching implications for telecommunications, computing, and data storage industries, driving progress in global connectivity and digital infrastructure.

Healthcare has seen transformative advancements through microgravity-enabled technologies, particularly in pharmaceuticals and tissue engineering. In space, protein crystallization occurs with extraordinary precision, enabling scientists to map protein structures with high resolution. This capability has been instrumental in developing more effective drugs, particularly for diseases such as cancer, Alzheimer's, and Parkinson's. Additionally, microgravity offers an optimal environment for growing tissue scaffolds and bioengineered organs. On Earth, gravitational forces can cause uneven cell growth and scaffold deformation, but in microgravity, cells grow uniformly and develop more naturally. These advancements hold promise for regenerative medicine, personalized therapies, and organ transplantation, potentially reducing the reliance on organ donation and mitigating the challenges of transplant rejection.

The energy sector has also benefited from microgravity research, particularly in the design of advanced energy storage systems and insulation materials. Batteries and supercapacitors manufactured in microgravity exhibit more uniform material deposition, resulting in increased energy density, longer lifespans, and improved efficiency. These developments are crucial for renewable energy systems, portable electronics, and electric vehicles. Similarly, insulation materials designed for space environments offer exceptional thermal resistance and durability, making them ideal for applications in both terrestrial and space-based systems. Improved heat transfer mechanisms, developed through microgravity experiments, contribute to more efficient thermal management in energy production and storage facilities, further reducing energy waste and operational costs.

Environmental science has also reaped significant rewards from microgravity research, particularly in the fields of water and air purification. Space missions require highly efficient life-support systems capable of recycling water and purifying air in closed-loop environments. Technologies developed for these missions, including advanced filtration membranes and catalytic converters, have been adapted for use on Earth to address water scarcity and pollution challenges. Microgravity-enabled filtration systems can remove microscopic contaminants more effectively than traditional methods, providing clean drinking water to remote or disaster-stricken areas. Similarly, air purification technologies developed for spacecraft are being repurposed to reduce indoor air pollution and improve air quality in industrial and residential spaces.

The cross-industry technological impacts of microgravity research demonstrate its versatility and potential to address some of the most pressing challenges facing humanity. In aerospace, lighter and more durable materials are enhancing spacecraft capabilities, while in the automotive sector, fuel-efficient engines and lightweight components are reducing emissions and operational costs. In electronics, ultrapure semiconductors and advanced optical fibers are revolutionizing global communication and computing networks. Healthcare is witnessing breakthroughs in drug development and tissue engineering, offering new hope for treating previously incurable diseases. The energy sector benefits from highly efficient storage and insulation systems, advancing the global transition to renewable energy sources. Meanwhile, environmental science leverages space-based filtration and purification technologies to tackle water scarcity and air pollution on Earth.

Microgravity's unique ability to eliminate gravitational biases in physical and chemical processes makes it a critical tool for pushing the boundaries of scientific discovery and technological advancement. For instance, experiments in fluid dynamics conducted aboard space stations have led to improvements in terrestrial mixing processes, fuel efficiency, and thermal management. Similarly, heat transfer technologies developed for spacecraft are now being employed in industrial cooling systems, enhancing performance and reducing energy consumption. These cross-sectoral synergies demonstrate that innovations originating from microgravity research have applications far beyond their initial experimental scope.

Private space companies, alongside government agencies, are playing a pivotal role in expanding access to microgravity environments through orbital and suborbital platforms. As commercial spaceflight becomes more accessible, industries will have greater opportunities to leverage microgravity for research and development. This democratization of space-based experimentation promises to accelerate innovation across sectors, creating new markets for microgravity-enabled products and technologies. Additionally, the knowledge gained from space research is informing strategies for sustainable resource management and environmental conservation, addressing global challenges with science-based solutions.

The future of microgravity research holds immense promise for transformative advancements in industries ranging from aerospace and healthcare to energy and environmental science. Emerging fields such as space-based biomanufacturing, self-assembling nanomaterials, and advanced composites are set to revolutionize technological capabilities on Earth. Furthermore, the integration of AI and ML into space experiments is expected to optimize research outcomes, accelerating the development of space-enabled technologies. These innovations are not only enhancing life on Earth but are also laying the foundation for sustainable human presence in space, with implications for long-duration space missions and planetary colonization.

Microgravity represents an invaluable environment for driving technological advancements across aerospace, automotive, electronics, healthcare, energy, and environmental science sectors. Each industry benefits from the unique conditions provided by space, including enhanced material properties, precision manufacturing, optimized energy systems, and innovative environmental technologies. As humanity continues to expand its presence in space through international collaboration and commercial ventures, microgravity will remain a cornerstone for scientific discovery and industrial innovation. Technologies developed in space not only address current global challenges but also pave the way for a sustainable and technologically advanced future on Earth and beyond.

FUTURE DIRECTIONS IN MICROGRAVITY RESEARCH

- Space-Based Manufacturing Facilities: Dedicated modules for continuous material and pharmaceutical production in orbit.
- Deep Space Exploration: Materials optimized for extreme conditions on Mars and the Moon.
- Commercial Partnerships: Collaboration between space agencies and private industries for technology transfer.
- Sustainable Technologies: Leveraging microgravity findings to address climate change and resource scarcity on Earth.

Microgravity research serves as a catalyst for innovation across industries, enabling advancements in aerospace, automotive, healthcare, electronics, energy, and environmental sectors. These insights not only address challenges in space exploration but also drive groundbreaking technologies and sustainable solutions on Earth. As commercial space activities expand, the integration of microgravity-enabled technologies into everyday life will become increasingly significant.

Microgravity research stands at the forefront of scientific and technological innovation, offering unique opportunities to address some of humanity's most complex challenges. As space technology advances and access to LEO expands, the next logical step involves establishing space-based manufacturing facilities dedicated modules designed to produce advanced materials, pharmaceuticals, and other high-value products in orbit. These facilities would provide uninterrupted access to microgravity conditions, allowing for the large-scale production of materials with properties unattainable on Earth. For instance, high-purity semiconductors and defect-free crystals, crucial for advanced electronics and pharmaceutical drug formulations, could be produced with consistent quality. Space-based manufacturing also holds the potential to create specialized optical fibers and nanomaterials, both of which rely on precise environmental conditions during production. By transitioning these processes to space-based platforms, industries could significantly reduce costs associated with experimentation and improve scalability, eventually making space-manufactured products commercially viable on Earth.

Another promising frontier in microgravity research is its application to deep space exploration, where optimized materials and technologies are essential for survival and operational success on the Moon, Mars, and beyond. Materials manufactured in microgravity exhibit superior structural integrity, reduced defects, and enhanced thermal properties, making them ideal for spacecraft, habitats, and energy systems required for long-duration missions. For example, heat-resistant alloys, lightweight composites, and durable polymers developed in microgravity can withstand the harsh temperature fluctuations and radiation exposure encountered in deep space environments. Additionally, biological experiments conducted in space allow scientists to better understand how human bodies respond to microgravity over extended periods, informing medical interventions and physical training regimens for astronauts. These findings are vital for planning sustained missions on extraterrestrial surfaces, ensuring both safety and efficiency in resource management, habitat construction, and overall mission design.

In parallel, commercial partnerships between space agencies and private industries are becoming a cornerstone of microgravity research, driving innovation through collaboration and technology transfer. As private space companies develop reusable rockets and deploy modular space stations, the cost of accessing microgravity environments is expected to decline, opening doors for smaller businesses and research institutions to participate. Public–private partnerships are enabling more industries to leverage microgravity platforms for experimentation and product development, leading to a new wave of space-enabled technologies. For example, pharmaceutical companies are increasingly investing in protein crystallization studies aboard space stations to develop more effective drugs. Similarly, telecommunications companies are exploring the manufacturing of defect-free optical fibers in orbit to improve global connectivity. These collaborations create a feedback loop of innovation, where terrestrial technologies benefit from space-based insights and space missions gain from Earth-bound expertise.

Microgravity research also offers critical insights into the development of sustainable technologies, helping address pressing challenges such as climate change, resource scarcity, and environmental degradation. Innovations developed in space environments, including water purification systems, advanced insulation materials, and energy-efficient technologies, are finding terrestrial applications in resource management and sustainability projects. For example, water recycling systems originally designed for space habitats are now being adapted for use in arid regions and disaster-stricken areas, providing clean drinking water where traditional infrastructure is unavailable. Likewise, thermal insulation technologies engineered for spacecraft are being repurposed for energy-efficient buildings and industrial facilities, reducing heating and cooling demands. Microgravity experiments also provide valuable data for optimizing renewable energy systems, such as solar panels and next-generation energy storage devices, which play a critical role in reducing reliance on fossil fuels.

In the future, space-based pharmaceutical production is expected to become a reality, with dedicated orbiting labs focusing on manufacturing complex medications that require highly controlled crystallization or molecular assembly processes. Protein crystallization experiments in microgravity have already demonstrated their potential to unlock new treatments for diseases such as cancer, Alzheimer's, and diabetes. In the coming years, dedicated manufacturing platforms could mass produce these specialized drugs, lowering costs and improving global accessibility. Furthermore, biotechnology research in microgravity may lead to advancements in personalized medicine, where treatments can be fine-tuned at the molecular level to match an individual patient's biology.

The establishment of orbital research hubs will allow for long-term, uninterrupted studies in fields ranging from material science to biotechnology. Unlike current space stations, which balance a wide range of scientific and operational objectives, future orbital hubs may be specialized for specific industries, such as pharmaceuticals, energy, or semiconductors. These platforms would reduce the logistical bottlenecks currently associated with short-duration experiments, enabling continuous manufacturing and observation cycles. For example, materials that require weeks or months of microgravity exposure to fully develop could be produced without interruption, enhancing both efficiency and quality. Such hubs would also serve as training grounds for future space scientists and engineers, preparing the next generation for careers in space-based research and manufacturing.

One of the most compelling directions for microgravity research lies in the development of self-sustaining space habitats capable of supporting long-term human presence in orbit or on other celestial bodies. Lessons learned from managing closed-loop life support systems on spacecraft are directly applicable to sustainable living on Earth, particularly in urban environments facing resource scarcity. Water recycling, air purification, and waste management systems originally designed for space are being adapted for smart cities, enabling more efficient resource use and reducing environmental footprints. Furthermore, agricultural experiments conducted in microgravity, such as vertical farming techniques and soil-free crop growth, have significant implications for addressing global food insecurity. These insights could revolutionize terrestrial agriculture by improving crop yield, reducing water consumption, and minimizing reliance on chemical fertilizers.

AI and robotics are also expected to play a crucial role in the future of microgravity research. AI-driven analytics will enable scientists to process vast amounts of data collected from space experiments more efficiently, identifying patterns and insights that would otherwise go unnoticed. Autonomous robots will handle complex manufacturing tasks aboard space stations, reducing the need for human intervention and minimizing risks associated with prolonged space exposure. Together, AI and robotics will streamline space-based production, optimize experimental protocols, and ensure consistent quality control in orbit.

Furthermore, the future of microgravity research will likely see an increase in educational and outreach initiatives aimed at inspiring the next generation of scientists, engineers, and entrepreneurs. Space agencies and commercial companies are already launching educational programs that allow students to design and conduct their own microgravity experiments aboard orbital platforms. These initiatives not only foster a passion for STEM education but also cultivate a workforce equipped with the skills necessary for the emerging space economy.

The future of microgravity research promises groundbreaking advancements across diverse fields, including space-based manufacturing, deep space exploration, commercial partnerships, and sustainable technologies. By establishing dedicated orbital facilities, optimizing materials for extraterrestrial environments, fostering public–private collaborations, and addressing global sustainability challenges, humanity is poised to unlock unprecedented opportunities. The integration of AI, robotics, and advanced biotechnology will further accelerate these developments, ensuring that space remains a hub of innovation. As access to microgravity becomes more widespread, the ripple effects of space-based discoveries will continue to transform industries, improve quality of life on Earth, and pave the way for humanity's long-term presence beyond our planet.

Bibliography

Andraos, J., & Matlack, A. S. (2022). *Introduction to green chemistry*. CRC Press.

Bolonkin, A. (2007). Micro-thermonuclear AB-reactors for aerospace. *Aircraft Engineering and Aerospace Technology*, *79*(6), 646–665. https://doi.org/10.1108/00022660710829845

D'Angelo, O., Kuthe, F., Liu, S.-J., Wiedey, R., Bennett, J. M., Meisnar, M., Barnes, A., Kranz, W. T., Voigtmann, T., & Meyer, A. (2021). A gravity-independent powder-based additive manufacturing process tailored for space applications. *Additive Manufacturing*, *47*, 102349. https://doi.org/10.1016/j.addma.2021.102349

Dunn, B. D. (2016). *Materials and processes: For spacecraft and high reliability applications*. Springer. https://doi.org/10.1007/978-3-319-23362-8

Ehrfeld, W., Hessel, V., & Löwe, H. (Eds.). (2000). *Microreactors: New technology for modern chemistry*. Wiley-VCH.

Elvis, M. (2021). *Space resources: Physical constraints, policy choices, & ethical considerations*. MIT Press.

Hessel, V. (Ed.). (2018). *Design and engineering of microreactor and smart-scaled flow processes*. Elsevier.

Hessel, V., Stoudemire, J., Miyamoto, H., & Fisk, I. D. (Eds.). (2022). *In-space manufacturing and resources: Earth and planetary exploration applications*. Wiley-VCH.

Ilic, O. (2020). Nanophotonic materials for space applications. *MRS Bulletin*, *45*(9), 769–778. https://doi.org/10.1557/mrs.2020.223

Knowles, J. P., Elliott, L. D., & Booker-Milburn, K. I. (2008). *Micro-photochemistry: Photochemistry in microstructured reactors*. Wiley-VCH.

Lappa, M. (Ed.). (2004). *Fluids, materials and microgravity: Numerical techniques and insights into the physics*. Springer. https://doi.org/10.1007/b139019

Mohamed, S. E. (2023). *Nuclear propulsion and space microreactors*. [Publisher not specified; likely academic press or technical report].

Myshkis, A. D. (1987). *Low-gravity process engineering*. Springer US.

O'Neill, G. K. (Ed.). (1980). *Space-based manufacturing from nonterrestrial materials*. Science Research Associates.

Patel, S. (2023). What are microreactors and how soon could we see one in operation? *Chemical & Engineering News*. American Chemical Society. [Exact issue/date not retrieved].

Paton, B. (Ed.). (2003). *Space technologies, materials and structures*. Taylor & Francis.

Sciti, D., Zoli, L., & Vinci, A. (Eds.). (2022). Ultra-high temperature ceramic matrix composites for aerospace applications. *Composites Part B: Engineering, 234*, 109709. https://doi.org/10.1016/j.compositesb.2022.109709

Sivolella, D. (2019). *Space mining and manufacturing: Off-world resources and revolutionary engineering techniques*. Springer. https://doi.org/10.1007/978-3-030-30881-0

Tanaka, Y. (2002). Microreactors: Prospects already achieved and possible misuse. [Journal and DOI not available].

Wirth, T. (Ed.). (2013). *Microreactors in organic chemistry and catalysis* (2nd ed.). Wiley-VCH.

Glossary

Additive Manufacturing (AM) A process of creating objects layer by layer using digital models, commonly known as 3D printing. In space, AM allows the production of tools, spare parts, and materials using in-situ resources.

Additive Manufacturing A process of creating three-dimensional objects layer by layer, commonly known as 3D printing. In microgravity, it allows precise control over material deposition without gravitational distortion.

Alloy Solidification The process by which molten alloys transition into solid structures. In microgravity, solidification occurs without gravitational segregation, producing uniform microstructures.

Alloy A mixture of two or more metals, or a metal and a non-metal, designed to exhibit enhanced mechanical, thermal, or chemical properties.

Amorphous Metals Metals that lack a long-range atomic order, also known as metallic glasses. They are often synthesized in microgravity due to slower cooling rates and uniform heat dissipation.

Autonomous Laboratory Systems AI and robotics-enabled systems used in space to perform material synthesis experiments with minimal human intervention.

Autonomous Microreactor System A self-sustaining microreactor equipped with automated control systems and AI algorithms to operate without human intervention, essential for long-term missions.

Biocompatible Polymers Polymers that can interact with biological systems without causing adverse reactions. They are used in medical implants, tissue scaffolding, and drug delivery systems in space habitats.

Biodegradable Polymers Polymers designed to break down naturally in biological environments. In microgravity, their synthesis results in improved structural uniformity and predictable degradation rates.

Biomedical Devices Tools and implants created from advanced materials, often using microreactors, to monitor astronaut health or deliver treatments during space missions.

Bulk Metallic Glass (BMG) A type of amorphous metal characterized by a disordered atomic structure, synthesized more effectively in microgravity environments.

Buoyancy-Driven Convection Fluid motion caused by density differences resulting from temperature or concentration gradients. In microgravity, convection is significantly reduced, allowing diffusion-dominated processes to prevail.

Catalysis A chemical process where a catalyst accelerates a reaction without being consumed in the process. Catalysis is crucial in microreactors for efficient chemical synthesis in space.

Ceramic Matrix Composites (CMCs) Materials made of ceramic fibers embedded in a ceramic matrix, offering high thermal resistance, durability, and strength, ideal for spacecraft shielding and high-temperature environments.

Ceramic Sintering A process of compacting and forming ceramic materials using heat and pressure. In microgravity, sintering results in reduced porosity and structural defects.

Chemical Vapor Deposition (CVD) A process in which chemical reactions occur on a substrate's surface to produce thin films. In microgravity, it ensures uniform deposition without gravitational distortions.

Closed-Loop System A system where resources, such as reagents, water, or air, are continuously recycled and reused, minimizing waste in space habitats.

Conductive Polymers Polymers that conduct electricity, used in flexible electronics, wearable sensors, and photovoltaic cells for space systems.

Containerless Processing Techniques for handling molten materials in microgravity without using physical containers, preventing contamination and enhancing purity.

Crystal Nucleation The initial step in crystal formation where atoms or molecules arrange into a defined, repeating pattern. Microgravity allows for uniform nucleation due to minimized convection.

Crystallization A process where atoms or molecules organize into a highly structured lattice, forming crystals. In microgravity, crystallization occurs with fewer defects and better uniformity.

Crystallization The process by which atoms or molecules form a structured lattice. In microgravity, crystallization occurs more evenly, reducing defects and inclusions.

Density Gradients Variations in density within a material caused by gravitational effects. In microgravity, these gradients are minimized, leading to homogeneity in synthesized materials.

Density-Driven Separation A phenomenon where heavier materials sink and lighter materials rise under gravity. This is eliminated in microgravity, ensuring homogeneous material synthesis.

Diffusion-Controlled Reactions Chemical reactions limited by the rate of molecular diffusion rather than reaction speed, which dominate in microgravity environments.

Diffusion-Dominated Processes Processes where mass transport occurs primarily through diffusion rather than convection. In microgravity, diffusion dominates material synthesis, resulting in uniform structures.

Drop Tower A terrestrial facility that creates short durations of microgravity by allowing materials to free-fall in a controlled chamber.

Elastomers Polymers with elastic properties, widely used in seals, gaskets, and flexible connectors in spacecraft systems.

Electrochemical Processes Chemical processes driven by electric current or voltage, used in microreactors for tasks like water splitting and fuel synthesis.

Electromagnetic Levitation (EML) A method for containerless processing of molten metals using electromagnetic fields to suspend the material, preventing contamination and gravitational effects.

Emulsion Polymerization A process of creating polymers in an emulsion, often stabilized by surfactants. In microgravity, it results in uniform droplet distribution and controlled polymer growth.

Enhanced Diffusion The predominant transport mechanism in microgravity, where molecular motion occurs uniformly without gravitational interference.

Extraterrestrial Resource Utilization (ISRU) The practice of collecting and processing local planetary or lunar resources (e.g., Martian CO_2 or lunar regolith) to produce water, oxygen, fuel, and building materials.

Fault Detection and Recovery (FDR) Systems embedded in microreactors to identify malfunctions and initiate corrective measures autonomously.

Fluid Dynamics in Microgravity The study of fluid behavior in low-gravity environments, where surface tension and diffusion dominate over buoyancy-driven convection.

Fuel Cells Electrochemical cells that convert chemical energy into electricity, commonly used in spacecraft power systems.

Functionally Graded Materials (FGMs) Materials with spatial variations in composition or structure, optimized for specific performance requirements. Microgravity allows seamless gradient transitions in FGMs.

Gravitational Segregation The separation of material phases or elements based on differences in density caused by gravity. Microgravity eliminates this phenomenon, ensuring homogeneous material synthesis.

Heat Transfer in Microgravity The movement of thermal energy within a material, occurring primarily via conduction and radiation rather than convection.

High-Entropy Alloys (HEAs) Metallic alloys containing five or more principal elements in near-equal proportions, offering exceptional mechanical strength, thermal stability, and corrosion resistance.

High-Strength Elastomers Polymers with elastic and high mechanical strength properties, used for shock absorption and sealing components in space habitats.

Hydrogel A network of polymeric materials with high water content, often used in drug delivery systems and tissue scaffolding in space medicine.

In-Situ Resource Utilization (ISRU) The process of harnessing and converting raw materials found in extraterrestrial environments such as lunar regolith, Martian atmospheric gases, or subsurface ice into usable resources like fuel, oxygen, water, and construction materials to support space missions, manufacturing, and sustained human presence beyond Earth.

International Space Station (ISS) A space-based laboratory in low-Earth orbit (LEO) designed for long-duration microgravity research, serving as a hub for scientific experiments in fields such as materials synthesis, microreactor studies, biology, and technology development in space environments.

Kinetics in Microgravity The study of reaction rates and molecular interactions in material synthesis processes under microgravity conditions, often dominated by diffusion rather than convection.

Levitation Techniques Methods (e.g., electromagnetic and electrostatic levitation) used in microgravity to suspend molten materials for containerless processing.

Lightweight Metallic Composites Metal-based materials designed to maximize strength while minimizing weight, crucial for spacecraft frames and cryogenic fuel tanks.

Long-Duration Microgravity Experiments Experiments conducted over extended periods aboard platforms like the ISS to study slow material synthesis processes.

Long-Term Sustainability The ability to maintain continuous operations in space through recycling, waste management, and resource optimization.

Material Homogeneity Uniform distribution of atoms or particles in a synthesized material, often achieved in microgravity due to reduced convection and sedimentation.

Microfluidics The science of controlling small volumes of fluids through microscopic channels, integral to microreactor technology for precise chemical reactions.

Microgravity An environment where gravitational forces are significantly reduced, often described as weightlessness. This condition enables unique physical phenomena such as uniform mixing, precise crystallization, and enhanced material synthesis that are difficult or impossible to achieve under Earth's gravity.

Microreactor A miniaturized chemical reactor capable of conducting chemical reactions with high precision, efficiency, and reduced resource consumption.

Modular Reactor Design A reactor design strategy where components are interchangeable and easily replaceable, allowing quick repairs and scalability.

Nanocomposites Materials composed of nanoparticles dispersed in a matrix, synthesized in microgravity for improved dispersion and uniform properties.

Nanostructures Engineered materials at the nanometer scale that exhibit unique physical, chemical, and optical properties. They are designed to manipulate light and energy, making them essential in applications such as optical sensors, photonic circuits, high-speed communication devices, and advanced material systems often benefiting from synthesis in microgravity environments.

Nucleation Sites Locations where crystal growth begins. In microgravity, nucleation occurs more uniformly due to reduced convection.

Nucleation The initial process of crystal or polymer formation. In microgravity, nucleation occurs uniformly, reducing defects in the final material.

Optical-Grade Materials Materials synthesized with high transparency, uniform refractive indices, and minimal defects, used in laser systems and telescope optics.

Parabolic Flight A technique for generating short-duration microgravity environments using specially equipped aircraft performing parabolic maneuvers.

Phase Separation The formation of distinct regions in a material due to differences in component properties. In microgravity, phase separation occurs uniformly without gravitational distortions.

Photonic Crystals Nanostructures designed to control and manipulate light, used in optical sensors, photonic circuits, and high-speed communication devices.

Polymerization A chemical reaction in which small molecules, called monomers, bond together to form long polymer chains. In microgravity, this process benefits from uniform chain growth, improved control over molecular structure, and reduced structural defects, enabling the creation of higher-quality polymers.

Radiation Shielding Materials Materials designed to absorb and deflect harmful radiation, often synthesized in microgravity to enhance protective properties.

Reagent Recycling The practice of recovering and reusing chemical reagents in closed-loop systems aboard spacecraft.

Reduced Gravity Conditions where gravitational forces are significantly minimized, as experienced in orbit or during free-fall in drop towers.

Round-Robin Testing A method of verifying material properties through repeated testing under varying conditions, often applied to microgravity experiments.

Scalability The ability to expand microreactor operations for larger-scale material production in space.

Sedimentation The gravitational settling of denser particles within a liquid or slurry. In microgravity, sedimentation is eliminated, resulting in uniform particle distribution.

Self-Healing Materials Advanced materials capable of autonomously repairing damage, crucial for long-term missions and space habitats.

Shape-Memory Alloys (SMAs) Metal alloys that return to their original shape when heated, used in self-actuating spacecraft components.

Superalloys High-performance metallic alloys designed to withstand extreme temperatures and stress, often synthesized in microgravity for uniform microstructures.

Telemetry The remote collection and transmission of data from micro-reactors to Earth-based monitoring systems.

Thermal Gradients Differences in temperature within a material. In microgravity, these gradients are stabilized, ensuring uniform heat distribution.

Thermal Insulation Polymers Polymers designed to minimize heat transfer, used in spacecraft thermal protection systems.

Thermodynamics in Microgravity The study of energy transfer and phase equilibria under low-gravity conditions.

Vacuum Environment The near-absence of matter in space, influencing chemical reactions, crystallization processes, and material properties.

Vapor Deposition Techniques for producing thin material films by depositing vaporized material on a substrate. Microgravity ensures uniform coating without gravitational interference.

Waste Management Systems Technologies for handling, recycling, and neutralizing chemical and biological waste aboard spacecraft.

Weightlessness The sensation or condition experienced in microgravity, where gravitational forces are near zero.

Zero Gravity (0g) A state where the effects of gravity are negligible, typically achieved in orbiting spacecraft or drop towers.

Further Reading

Exploring the synthesis of materials in microgravity environments offers unique insights into the role of gravity in material formation and the potential for creating novel materials with enhanced properties. Below is a curated list of books that delve into various aspects of this field, providing foundational knowledge and discussing advancements in microgravity research.

Fluids, Materials and Microgravity
Authored by Marcello Lappa, this book examines the behavior of fluids and materials in microgravity, emphasizing numerical techniques and physical insights. It provides a comprehensive understanding of how microgravity affects material properties and processing.

Handbook on Synthesis Strategies for Advanced Materials
Edited by A. K. Tyagi and Raghumani S. Ningthoujam, this handbook presents state-of-the-art coverage of unconventional synthetic routes for advanced functional materials, including those synthesized under microgravity conditions. It discusses various synthesis methods and their applications.

Physical Science Under Microgravity: Experiments on Board the SJ-10 Recoverable Satellite
This book provides a comprehensive review of experiments conducted in microgravity, focusing on fluid physics and materials science. It highlights the unique physical and chemical phenomena observed in microgravity environments.

Materials Syntheses: A Practical Guide

Authored by Ulrich Schubert, Nicola Hüsing, and Richard Laine, this guide addresses the complexities of material synthesis, emphasizing reproducibility and clarity in protocols. While not exclusively focused on microgravity, it provides foundational knowledge applicable to understanding material behavior in different environments.

The World of Materials

This introductory materials science book covers structure, synthesis, properties, and modern-day material challenges. It offers a broad survey of materials science, including discussions relevant to microgravity research.

These books provide a comprehensive understanding of the challenges and opportunities in material synthesis, both on Earth and in microgravity environments. They serve as valuable resources for researchers and students interested in the field of materials science and engineering.

The following books provide comprehensive insights into these fundamental effects and their implications for material behavior in microgravity:

Physical Science Under Microgravity: Experiments on Board the SJ-10 Recoverable Satellite

Edited by Wenrui Hu and Qi Kang, this book presents experiments conducted in a space microgravity environment, focusing on fluid physics and materials science. It offers insights into the kinetic properties of matter under microgravity conditions.

Thermodynamics and Kinetics in Materials Science: A Short Course

This book discusses the basic laws of classical and irreversible thermodynamics, phase equilibria, theory of solutions, and chemical reaction kinetics, providing a solid foundation for understanding material behavior in varying environments, including microgravity.

An Introduction to Aspects of Thermodynamics and Kinetics Relevant to Materials Science

Authored by Eugene Machlin, this book applies thermodynamics and kinetics to solve problems within materials science, offering insights that are pertinent to understanding material synthesis in microgravity.

Introduction to Thermodynamics and Kinetic Theory of Matter

This book deals with areas of molecular physics, covering elementary statistical mechanics and kinetic theory, and includes thermodynamics, providing innovative methods for calculating surface structure within quasi-thermodynamics.

These resources offer valuable theoretical frameworks and experimental insights into the fundamental effects of microgravity on material behavior, thermodynamics, and kinetics, essential for researchers and students in the field of materials science and engineering.

Index

FSC
www.fsc.org
MIX
Papier | Fördert gute Waldnutzung
FSC® C083411

Kolibri 360 GmbH
Ferdinand-Jühlke-Straße 7
99095 Erfurt, Deutschland
produktsicherheit@kolibri360.de